SpringerBriefs in Computer Science

SpringerBriefs present concise summaries of cutting-edge research and practical applications across a wide spectrum of fields. Featuring compact volumes of 50 to 125 pages, the series covers a range of content from professional to academic.

Typical topics might include:

- A timely report of state-of-the art analytical techniques
- A bridge between new research results, as published in journal articles, and a contextual literature review
- A snapshot of a hot or emerging topic
- An in-depth case study or clinical example
- A presentation of core concepts that students must understand in order to make independent contributions

Briefs allow authors to present their ideas and readers to absorb them with minimal time investment. Briefs will be published as part of Springer's eBook collection, with millions of users worldwide. In addition, Briefs will be available for individual print and electronic purchase. Briefs are characterized by fast, global electronic dissemination, standard publishing contracts, easy-to-use manuscript preparation and formatting guidelines, and expedited production schedules. We aim for publication 8–12 weeks after acceptance. Both solicited and unsolicited manuscripts are considered for publication in this series.

**Indexing: This series is indexed in Scopus, Ei-Compendex, and zbMATH **

Yanlong Zhai • Muhammad Mudassar •
Liehuang Zhu

Edge Computing Resilience

Overcoming Resource Constraints
in Unstable Computing Environments

 Springer

Yanlong Zhai (iD)
Cyberspace Science and Technology
Beijing Institute of Technology
Beijing, China

Muhammad Mudassar (iD)
Computer Science Department
COMSATS University Islamabad
Vehari, Punjab, Pakistan

Liehuang Zhu (iD)
Cyberspace Science and Technology
Beijing Institute of Technology
Beijing, China

ISSN 2191-5768   ISSN 2191-5776   (electronic)
SpringerBriefs in Computer Science
ISBN 978-981-97-6997-1   ISBN 978-981-97-6998-8   (eBook)
https://doi.org/10.1007/978-981-97-6998-8

This Springer imprint is published by the registered company Springer Nature Singapore Pte Ltd.
The registered company address is: 152 Beach Road, #21-01/04 Gateway East, Singapore 189721,
Singapore

If disposing of this product, please recycle the paper.

To the visionaries and innovators in the domain of edge computing, whose tireless quest for excellence and a determined focus on forging ahead in technology have motivated us to venture into new horizons of computing resilience.

This book pays tribute to those who demonstrate anticipatory wisdom to identify potential amidst difficulties, the engineers who persistently refine resource utilization, and the researchers who boldly address the ever-shifting landscape, all in gratitude for your unswerving commitment and passion.

We also extend our heartfelt appreciation to our families, whose steady solidarity and deep concern have sustained us on this voyage. With your love and support, we've found the strength to fulfill ambitious objectives.

May this book light the way for those enthusiastic about enhancing the resilience, efficiency, and adaptability of computing environments, serving as a source of knowledge and inspiration.

Foreword

In an era of growing global interconnectivity, the essence of technological innovation is changing at its periphery. Edge computing, a forward-thinking notion that spurring readily available infrastructure, is altering our perspective of data processing, scalability, and fault tolerance, and it's increasing the potential for innovation.

"Edge Computing Resilience: Overcoming Resource Constraints in Unstable Computing Environments" demonstrates the unwavering commitment to pursue knowledge and distinguish the relevant research and industry sectors in this dynamically changing sector. Dr. Yanlong Zhai and a dedicated team of experts have scripted this book, thoroughly examining edge computing that spans from its foundational principles to progressive approaches. It presents valuable insights for those tracing the complex roadmap of edge systems. In this book, readers will discover a rich source of information that probes into the complicated world of edge computing from multiple angles.

Beginning with an introduction to edge computing systems, the monograph describes their real-world usability and the challenges they confront. The authors then focus on creating resilient edge systems, highlighting scalability and fault tolerance as essential aspects. By exploring resource-constrained offloading, energy-aware offloading, and optimization methodologies in subsequent chapters, readers will acquire essential knowledge to deal with these specific edge computing challenges. Looking ahead, the book also scrutinizes forthcoming ventures, staging a glance at the emerging cooperative edge computing models, wireless environments, and atypical scenarios.

In this masterwork, Dr. Yanlong Zhai's leadership and the authors' accumulated expertise glow brightly. This book reflects their commitment to establishing new paths in the domain of edge computing. It functions as a navigational tool for researchers, practitioners, and students embarking on their journey into this field. Whether you're an experienced professional looking to enhance your knowledge or a newcomer determined to explore this dynamic field, you will find invaluable enlightenment throughout the content.

Congratulations to Dr. Yanlong Zhai and the team for this distinguished contribution to the realm of edge computing. May "Edge Computing Resilience: Overcoming Resource Constraints in Unstable Computing Environments" encourage and brighten the way ahead for all who read it.

University of Wollongong, Wollongong, NSW,   Professor Jun Shen
Australia
Beijing, China
June 2024

Preface

The recent development of Edge Computing resulted in a new age of opportunities and problems in the ever-evolving field of computing. This book explores the complex domain of edge computing systems and offers a thorough examination of its core ideas, uses, and approaches for boosting effectiveness. The journey starts with the foundational work by outlining the fundamental ideas behind edge computing. While highlighting the significance of resilience in edge computing settings, we investigate its broad range of applications and the distinctive problems it brings.

The essential domain of scalability and fault tolerance considered critical to real-time edge applications is explored. We outline the function of fault tolerance and analyze the different sources and fault kinds. We focus on scaling in distributed edge systems and bottleneck identification, which prompts us to investigate scaling solutions and state backup techniques. The relevance of offloading in edge computing is explained along with the relationship between local and offloading devices. With an emphasis on inductive learning and privacy constraints, we investigate privacy-preserving offloading techniques since privacy is an important concern.

Furthermore, we focus on energy optimization and efficiency. We examine the energy used by offloading devices and compare the execution of local and offloading tasks. New offloading optimization techniques have been presented to support both single and multiple battery-operated devices. A thorough investigation of optimization in edge computing systems is presented. We explore several optimization strategies, such as task offloading, edge caching, network optimization, resource allocation, and privacy-aware optimization. In addition, edge node clustering and heterogeneous device settings are examined. We firmly think that readers' research into the field of edge computing will provide them with new perspectives and sources of inspiration.

Overview of the Book

This book is aimed to focus on three major aspects of edge computing resilience. Specifically, this includes fault detection and tolerance in edge computing, offloading, and optimization techniques in edge computing. As depicted in the book title, the book provides a comprehensive analysis of resilience in edge computing applications based on the aforementioned key areas. The discussion includes a thorough analysis of edge computing challenges, problem formulation, design methodologies, and implementation details of some application scenarios. It also provides recommendations for overcoming resource constraints in unstable computing environments based on the existing system designs and the future scope. It's significant to point out that optimization techniques, offloading methods, and fault-handling principles provided in this book are discussed within the scope and context of the edge computing paradigm.

In total, this book contains seven chapters, with multiple sections and subsections in each. Chapter 1 introduces the background of edge computing systems, their applications, and the challenges which prompted the need for resilience in edge computing applications. Several other challenges including security, system and statistical heterogeneity, resource management, and disaster management are discussed briefly, while the detailed discussion on fault tolerance, offloading, and optimization continues in the later section. Particularly, fault tolerance is covered in Chap. 2, whereas Chaps. 3 and 4 discuss about resource-constrained offloading and privacy preserving offloading respectively. SDN-based and energy-aware offloading is presented in Chap. 5. The edge computing optimization is covered in Chap. 6 while Chap. 7 focuses on the future scope.

In Chap. 2, we introduce the concept of scalability and fault tolerance for real-time edge applications. Here, we discuss various aspects of fault tolerances including their significance, causes, and different fault types. We presented design methodology and provided implementation of scale-out operation for fault tolerance. In detail, we give an insight into scaling for distributed edge system and provide a detailed design and operation of scale-out methods and state backup methods for fault tolerance respectively. Under this, we presented the principle of dynamic scaling where the system scales out resources in real-time according to the fluctuations of incoming streams with minimal overhead and latency. Furthermore, we also provide a concept and framework for check-pointing-based fault tolerance with scale-out operations.

In Chap. 3, we explore resource-constrained offloading, and then discuss offloading strategies and possible scenarios. A computational model for local execution and task offloading is introduced. The chapter examines existing research on task dependency, offloading to fog devices, dynamic offloading frameworks, and centralized versus distributed task offloading. Finally, the challenges linked to resource-constrained offloading are highlighted. Besides introducing resource-constrained offloading in edge computing, this chapter serves as an important prerequisite for the next chapter.

In Chap. 4, we uncover the need for privacy-preserving offloading in resource-constrained environments. A privacy-preserving offloading scheme for mobile edge intelligent systems is also presented. In this context, we discuss the privacy issues associated with task offloading and then present local differential privacy and inductive learning as a solution to address privacy concerns in edge computing. Further discussion includes energy-aware offloading which is provided in the next chapter.

In Chap. 5, our discussion is mainly centered on energy-aware offloading based on SDN. Three key areas are covered in this chapter, namely, the energy consumed by offloading devices, offloading task execution including the local and offloading execution tasks, and finally the offloading optimization which presents a detailed implementation of offloading optimization both for single and multiple devices on battery.

In Chap. 6, we introduce the concept and goal optimization for edge computing applications and then provide an in-depth discussion on various optimization techniques. These include task offloading, network optimization techniques, resource allocation schemes, the concept of edge caching as well, and privacy-aware optimization. We also presented the idea of optimization for heterogeneous devices followed by edge node clustering mechanisms for distributed computing introduced earlier. This chapter ends with a discussion on performance analyses and evaluation metrics such as profiling tools, benchmarking, tracing, simulations, containerization, and edge analytics.

The last chapter of the book, Chap. 7, focuses on future work in edge computing systems. In particular, we introduce the various challenges in the new cooperative edge computing. In addition, we provide a detailed discussion on optimization in wireless edge computing environment and provide potential recommendation and potential research directions for future work. Other aspects of edge computing resilience including edge computing in denied, disconnected, intermittent, or limited (D-DIL) environments are also discussed.

Beijing, China   Yanlong Zhai
Vehari, Pakistan   Mudassar Muhammad
Beijing, China   Liehuang Zhu
June 2024

Acknowledgments

A heartfelt appreciation to those outstanding individuals whose unwavering commitment and support have played a central role in bringing the book, "Edge Computing Resilience: Overcoming Resource Constraints in Unstable Computing Environments," to realization.

In commanding our team, Dr. Yanlong Zhai has played an indispensable role, presenting determined leadership, visionary insight, and top-tier competence over the entirety of this book's progress. He has provided the project with a solid framework through his supervision and loyalty. In their esteemed capacities as team members, Dr. Mudassar Muhammad and Dr. Liehuang Zhu have made substantial contributions, leveraging their knowledge, wisdom, and viewpoints to influence the content and progression of this work. Their collaborative contributions have deepened our deliberations and strengthened the book's comprehensiveness.

We want to acknowledge the Beijing Institute of Technology, Beijing, PR China, for their dominant role in the writing of a significant portion of this book. We owe a debt of gratitude to the resources and research atmosphere during our work. Moreover, we want to express our appreciation for the contributions made at COMSATS University Islamabad (Vehari Campus), Pakistan. Their association provided us with the opportunity to collaborate and enrich our understanding.

We would like to extend our true appreciation to Jude Tchaye-Kondi, Sarwar Adil, and Manjang Ousman, whose enthusiastic involvement and steadfast assistance throughout the writing journey have been incalculable. Your unwavering devotion to research, thorough data analysis, and the particular formation of this manuscript have been of utmost importance in upholding precision and diligence.

Alongside our families and loved ones, we also acknowledge the significant influence of our mentors, colleagues, and the wider academic community. Their guidance, feedback, and inspiration have been instrumental in our progress.

This book is a result of the collaborative endeavor and dedication of the individuals mentioned above. We truly value your contributions, and we are confident that this book will prove to be a requisite resource for researchers, practitioners, and enthusiasts in the field of edge computing.

Contents

Acronyms

AB	Active Backup
AE	Auto Encoder
AI	Artificial Intelligence
AP	Access Point
CC	Cloud Computing
CDC	Centers for Disease Control
CDN	Content Delivery Network
CEC	Cooperative Edge Computing
CO	Computation Offloading
CP	Content Provider
CSP	Cloud Service Providers
D2D	Device-to-Device
DAG	Directed Acyclic Graph
D-DIL	Denied, Disconnected, Intermittent, or Limited
DL	Deep Learning
DNN	Deep Neural Network
DoS	Denial of Service
DP	Differential Privacy
DSPS	Scaling Distributed Stream Processing Systems
EC	Edge Computing
ECR	Edge Computing Resilience
EDTs	Emerging and Disruptive Technologies
eMBB	Enhanced Mobile Broadband
ETP	Effective Throughput Percentage
FC	Fully Connected
FL	Federated Learning
GPS	Global Positioning System
HAR	Human Activity Recognition
HBLB	Points to an algorithm that takes into account the battery levels of all devices
IaaS	Infrastructure as a Service

IoT	Internet of Things
IoV	Internet of Vehicles
ITS	Intelligent Transportation System
JDG	JSON-Data-Generator
KL	Kullback-Leibler
LDP	Local Differential Privacy
LFU	Least Frequently Used
LiDAR	Light Detection and Ranging
LRU	Least Recently Used
MCC	Mobile Cloud Computing
MDP	Markov Decision Process
MEC	Mobile Edge Computing
MIMO	Multi Input Multi Output
ML	Machine Learning
MSE	Mean Square Error
Paas	Platform as a Service
PCA	Principal Component Analysis
PKI	Public Key Infrastructure
QoS	Quality of Service
RFID	Radio-Frequency Identification
RL	Reinforcement Learning
RSU	Roadside Units
RUR	Randomized Units Response
Saas	Software as a Service
SAE	Sparse AutoEncoder
SDN	Software Defined Network
SE	Secondary Execution
SGD	Stochastic Gradient Descent
SM	State Manager
SMD	Smart Mobile Device
SOTA	State-of-the-Art
SP	Service Provider
SPS	Stream Processing System
UE	User Equipment
V2X	Vehicle-to-Everything
VM	Virtual Machine

Chapter 1
Edge Computing

Abstract The development of Edge Computing Systems has transformed the way we handle and manage data in the rapidly evolving computing world. This chapter explores the fundamental concepts of edge computing and provides readers a comprehensive knowledge about this leading-edge technology. We start our exploration by looking into the fundamental ideas and architecture of edge computing systems, setting the stage for getting a deeper understanding of their practical uses and the challenges that they face. Edge computing is impacting a number of industries by improving IoT capabilities and lowering latency in critical operations. This has encouraged us to look at its different uses and the obstacles that still need to be addressed before it can be widely used. This chapter further emphasizes the need for resilience in edge computing and the requirement for robust frameworks that can resist unanticipated disturbances. Readers will be well-prepared for a journey into the complex world pertaining to edge computing and its prospects to revolutionize the field of information technology by the end of this chapter.

Keywords Edge computing · Distributed computing · Edge computing security · Resilience · Latency reduction · Industry impact

1.1 Introduction to Edge Computing Systems

Edge computing is a revolutionary approach to computing that encompasses the act of managing and appraising data at or near the source of data generation, rather than transmitting all the data to a centralized data repository or cloud-based platform. It is a distributed computing model that delivers computation and data storage near the locality where it is required, which is typically at the edge of the network. The term "edge computing" has been the subject of much debate and discussion in research and literature, with numerous definitions and interpretations offered by scholars and experts in the field that can be interpreted as:

- A transformative computing model that brings together computing, storage, and networking resources that are situated near the user, either geographically or in terms of network distance, to deliver efficient and high-quality application services.

- A platform near to data source that consolidates critical capabilities involving computing, networking, storage, and applications is available to deliver intelligent edge services that fulfill the industry's essential demands for agility in areas, for instance, connectivity, business applications in real-time, data optimization, intelligent applications, as well as security and privacy.
- An inventive computing framework is in operation at the network's edge, based on the principle that downlink data stands for cloud services, uplink data embodies the Internet of Everything, and the edge establishes the network and computational infrastructure bridging the data source with the route to the cloud computing center.

In contrast to conventional cloud computing, where data is handled in centralized data centers, edge computing involves placing computing resources closer to the devices and sensors that generate the data. Therefore, it is capable of lessening the processing latency involved in transmitting data to a centralized data store or cloud. Consequently, faster data processing, reduced network bandwidth usage, and improved application performance can result. This also expedites real-time decision-making, which holds significant value in diverse applications such as autonomous vehicles, smart industries, and healthcare.

Since, edge computing has the capability to revolutionize the way to handle and analyze data through faster, more effective, and more secure computing capabilities. Therefore, it is a source of considerable interest and investment, and it will be fascinating to observe how it develops and transforms our digital landscape in the next years.

Cloud and Edge Computing: Relationship and Distinction
In the era before edge computing, traditional cloud computing relied on sending all data to centralized cloud servers over the network, introducing significant challenges in managing computing and storage demands. Over time, cloud computing has evolved into a powerful network service platform, incorporating state-of-the-art technologies like load balancing, parallel computing, network storage, virtualization, and more. However, the explosive growth of Internet of Things (IoT) devices has overwhelmed networks with data, highlighting the limitations of cloud computing in meeting the real-time needs of time-sensitive systems. Thereby, cloud computing exhibits notable shortcomings in terms of load management, real-time data transmission, bandwidth efficiency, energy consumption, and data security and privacy.

Edge computing isn't a complete alternative to cloud computing; its emergence doesn't signify the end of cloud computing. For a successful digital transformation in industries, it's requisite for edge and cloud computing to collaborate, complement one another, and develop in harmony across network infrastructure, business strategies, application deployments, and intelligence augmentation. Although, data processing at edge nodes, aggregating data in the cloud is necessary to enable

Table 1.1 Cloud and edge computing: Fundamental contrast

	Applicable environment	Computation mode	Bandwidth pressure	Real time response
Cloud	Global	Centralized and large-scale data processing	More	High
Edge	Local	Intelligence analysis and small scale data processing	Less	Low

comprehensive analysis and gain profound insights. Therefore, cloud computing retains its significance as a crucial component for advancing more intelligent devices. Detailed distinctions between cloud and edge computing can be found in Table 1.1.

Cloud computing may experience substantial congestion when tasked with managing the immense data flow from interconnected devices, particularly within the context of the IoTs. This is where edge computing becomes invaluable, as it takes on tasks within its purview, thereby lightening the load. However, it's important to indicate that edge computing has its processing limits. For a comprehensive data analysis, the processed data must be relayed back to the cloud. This collaborative interplay between cloud and edge computing is key to efficient data analysis and dissemination. Their operational strategy may involve harnessing cloud computing for extensive big data analysis and insights generation, followed by data transfer to the edge for further processing and execution.

As presented in Fig. 1.1, the smooth integration of cloud computing and edge computing is actively shaping various facets of our daily lives, encompassing domains such as smart manufacturing, energy management, privacy and security

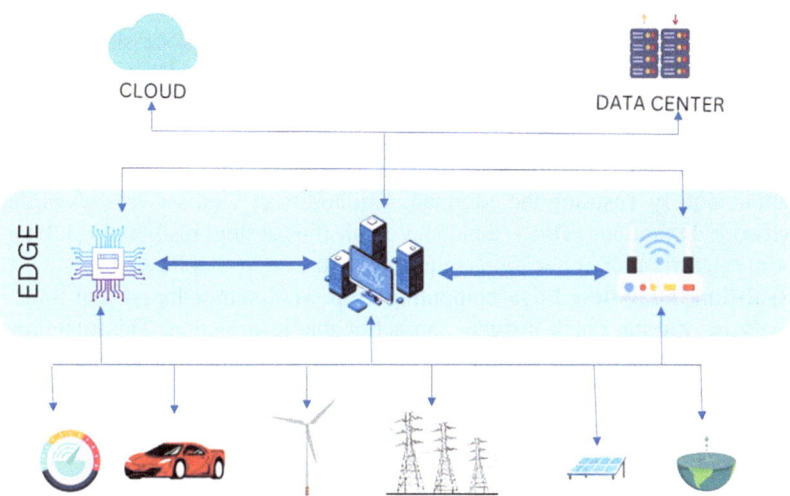

Fig. 1.1 Integrated environment of cloud and edge computing

enhancement, and home automation. Let's consider smart homes, as an example, where edge computing nodes encompass intelligent devices adept at processing heterogeneous data from a mass of sources. This processed data is then transmitted to the cloud for in-depth analysis, enabling cloud-based control of edge nodes and facilitating two-way communication. In the context of intelligent manufacturing, cloud computing offers centralized control, while edge nodes are equipped with real-time detection capabilities to swiftly address issues. Edge computing, renowned for its real-time capabilities, collaborates harmoniously with cloud computing to enhance production efficiency and promptly identify equipment anomalies. Addressing the requirements of IoT devices demands a purposeful consolidation of cloud computing and edge computing. It is through their joint expansion that the internet can evolve seamlessly.

1.2 Advantages of Edge Computing

Edge computing has a number of benefits over typical cloud computing. The most significant advantages are outlined here as:

- **Minimal delay and high throughput:** The strategic deployment of edge computing lowers data processing latency by positioning computation adjacent to the origin. Consequently, this accelerates response times and boosts the performance of real-time applications, such as IoT devices and autonomous vehicles.
- **Bandwidth efficiency:** Edge computing ejects the need for extensive data transfers to centralized cloud servers. Through on-site data processing, it diminishes the burden on network bandwidth, leading to a more cost-efficient and streamlined operation.
- **Enhanced data privacy and security:** By enabling local processing and storage of sensitive data, edge computing strengthens data breach defenses during data transit to distant data centers. This enhanced security and privacy is incredibly important in sectors such as healthcare and finance.
- **Resilience and redundancy:** Edge nodes have the capacity to function autonomously, ensuring the continual availability of vital services, even during network disruptions. This redundancy intensifies system resilience, making it a prime selection for applications requiring sustained availability.
- **Real-time analytics:** Edge computing empowers source-based real-time data analysis, offering quick insights and actionable information. This functionality is of critical significance for applications that rely on rapid decision-making, like predictive maintenance in manufacturing.
- **Scalability:** In edge computing, scalability can be easily attained by adding extra edge nodes as necessary. This adaptability empowers organizations to flexibly address shifting workloads and evolving needs, all without the need for substantial infrastructure modifications.

- **Lowered cloud costs:** By shifting selected processing tasks to the edge, businesses can adeptly lower their cloud computing expenses. The capability of edge devices allows them to filter and preprocess data, transmitting only the pertinent information to the cloud. This approach leads to cost reduction related to data transfer and storage.
- **Enhanced offline capability:** The ability of edge devices to operate autonomously, regardless of the lack of a connection to the central cloud infrastructure, is a beneficial attribute for applications that need to operate reliably in remote environments.
- **Compliance and regulatory benefits:** Organizations can achieve data sovereignty compliance through edge computing since it enables data processing and storage in specified geographic regions or legal jurisdictions, guaranteeing adherence to local legal requirements.
- **Edge AI and ML:** Edge computing expedites the deployment of artificial intelligence and machine learning models directly on edge devices, promoting immediate data analysis and real-time decision-making without the need for cloud-based resources
- **Reduced network congestion:** Data processing at the edge decreases the volume of unnecessary traffic toward centralized data centers, proficiently easing network bottlenecks. This, in turn, can lead to amplified network performance and decreased operational expenditures.
- **Improved disaster reinstatement:** Local backups granted by edge nodes evolve into a key resource, maintaining access to vital data and applications if a catastrophe occurs at central data centers.

1.3 Applications of Edge Computing

Edge computing is used in any application where data demands fast and immediate processing, and low latency is critical. The most common application domains of edge computing include:

- **IoT and smart devices:** Edge computing takes on a key role in IoT deployments by facilitating immediate data processing and analysis at the device's edge. This capably minimizes latency and accelerates on-spot decision-making. Consider the capacity of edge computing to empower smart home devices in processing sensor data autonomously, granting them the autonomy to manage lighting, thermostats, and security systems exempt from a central server.
- **Autonomous vehicles:** Edge computing forms the essence of self-driving cars, processing sensor data like LiDAR, cameras, and radar in real time. These edge nodes function as the intelligence behind prompt decisions regarding navigation, collision prevention, and traffic management, all intended to improve safety and efficiency.

- **Industrial automation:** Edge computing holds profound importance in industrial automation settings, providing the core foundation for low-latency control systems. It provides us with the resources needed for real-time machinery monitoring, preemptive maintenance, and ensuring uncompromising quality control standards.
- **Healthcare and telemedicine:** In healthcare, edge computing delivers its support to remote patient monitoring and telemedicine applications. Medical sensors and devices engage in the function of processing patient data at the edge, ensuring prompt alerts to healthcare practitioners while reducing the need for continuous connectivity. These advantages are particularly beneficial in remote or underserved localities.
- **Sales and consumer interaction:** Edge computing is an essential resource for retailers intending to individualize customer experiences. Through the real-time analysis of consumer behavior and preferences, stores can deliver adapted promotions and suggestions.
- **Video surveillance:** Edge computing is a vital part of video surveillance and security systems. Cameras possessing on-device video analysis capabilities can swiftly pinpoint suspicious activities or threats, causing quicker response times. It becomes especially significant in our efforts to protect critical infrastructure and ensure public safety.
- **Content dissemination and live streaming:** Edge servers are systematically embedded within Content Delivery Networks (CDNs) to overcome latency challenges in content delivery and streaming. Through locally caching content, edge computing ensures that video, gaming, and other media services offer smooth and uninterrupted streaming experiences.
- **Environmental monitoring:** Within environmental monitoring and conservation, edge computing is an essential resource. Sensors located in remote areas, such as forests or oceans, exhibit the potential to independently process data, recognizing shifts in temperature, humidity, pollution, or wildlife behavior. This data is then transmitted for a comprehensive evaluation.
- **Supply chain management:** Edge computing is an essential tool in the optimization of supply chain operations. Sensors and RFID tags on products and shipments allow real-time inventory tracking and condition monitoring. By offering critical information on the location, temperature, and humidity of goods, edge computing allows companies to make data-driven decisions and mitigate potential losses.
- **Crop management and precision farming:** Edge computing plays a key role in precision agriculture, enhancing crop management to its greatest potential. Within fields, sensors collect information on soil quality, moisture content, and contemporary weather conditions. Edge devices analyze this data to automate irrigation, fertilization, and pest management, which ultimately boost crop yields and sustainability.

1.4 Edge Computing Resilience (ECR)

As elaborated earlier in our discussion, edge computing emerges as a highly promising technology, facilitating the localization of computation and storage capabilities near data sources and users. This localization results in accelerated data processing diminished latency, and strengthened data confidentiality. However, edge computing systems frequently face various disruptions during their task execution phase, including network congestion, system failures (comprising hardware and software glitches), cyberattacks, and natural disasters. The origins of these disruptions can be traced predominantly to the following factors:

1. **Distributed and decentralized systems:** Unlike the traditional cloud-based centralized-based computing, the edge computing system is decentralized among multiple computing clients often distributed in different geographical areas. As a result, the system is fragile and prone to various kinds of disruption and faults.
2. **Resource constraints:** Many of the edge computing devices have low computational capabilities and there unable to execute highly complex and computationally intensive tasks. Similarly, energy consumption of is a crucial challenge to meet the growing demand of intensive operations.
3. **Heterogeneity:** Since EC operates in a distributed system, many of the computing devices possess unequal computing power which usually poses a challenge in large-scale application collaboration among end devices. Data heterogeneity posed similar threats. Therefore, it is of utmost importance to formulate and deploy resilience strategies within edge computing systems, safeguarding the persistent and trustworthy delivery of services via edge computing. **ECR** refers to the ability of an edge computing system to maintain its functionality and performance in the face of disruptions, failures, or attacks. Resilience is a crucial aspect of the design and operation of distributed computing systems at the edge of the network. The main goal of building resilience for edge computing systems is to enable the system to continue functioning in the face of disruptions, failures, or attacks. That is, edge devices are able to proceed with the normal task execution despite experiencing system constraints and malfunction.

Since the edge computing environment is decentralized among distributed computing nodes with no particular centralized authority, it is not feasible to avoid system failures which tend to disrupt normal operations. It is therefore logical to design and implement resilience which helps the system to adapt and cope with the challenge and provide efficient service to the end users.

To ensure an effective resilience system for IoT applications in an unstable computing environment, the following aspects are critical for design consideration:

1. **Fault tolerance:** Since, here resilience refers to a system's inherent capacity to persist in operation, even when faced with faults or failures. Fault tolerance mechanisms are skillfully designed to identify, isolate, and rectify these faults, thus curtailing their impact on the system's overall efficiency and availability. Within the context of edge computing, the integration of fault tolerance mecha-

nisms becomes predominant, fortifying the system against diverse disruptions. In the realm of edge computing, systems must possess the capability to withstand hardware or software failures within individual nodes while keeping the system's overall performance intact. This feat is accomplished through the tactical implementation of redundancy, replication, and load-balancing techniques.

Numerous fault tolerance mechanisms are applicable to edge computing systems, including redundancy, replication, load balancing as well as system monitoring and recovery:

- **Redundancy:** The redundancy is a common mechanism used to improve fault tolerance. By deploying redundant components, edge computing systems can continue operating even if one or more components fail. For example, using redundant servers, storage devices, or network connections can help secure the system's ongoing functionality in the event of a failure.
- **Replication:** Replication entails the generation of duplicates of data or services, distributing them across various locations and devices. By replicating data or services, edge computing systems can ensure that they are available even if a device or location fails.
- **Load balancing:** It is the practice of spreading workloads among numerous devices or locations to prevent any single device or location from becoming overwhelmed. By using load balancing, edge computing systems can improve fault tolerance by avoiding single points of failure and ensuring that workloads can be rerouted in the event of a failure.

2. **Offloading:** The strategic offloading of computational tasks is pivotal for ECR attainment. As edge computing aims to operate at the edge of the network, close to the data source, to minimize latency and improve data processing speed, edge devices have limited computing resources, which can lead to performance degradation or failure under heavy workloads or faults. Consequently, it is essential to provide a computational task offloading mechanism that builds resilience in the EC system by mitigating the impact of resource constraints and faults.

Computational task offloading involves transferring computation and data analysis tasks from edge devices to more powerful remote servers or cloud data centers in order to avoid overloading and enhance devices' performance. Computational task offloading can also be achieved by distributing computing tasks across multiple edge devices or servers, which can help ensure fault tolerance and reduce the impact of localized disruptions. By distributing tasks across multiple devices or servers, edge computing systems can ensure that processing tasks can continue even if one device or server fails.

In order to achieve ECR through computational task offloading, several factors need to be considered, such as workload characteristics, network conditions, security, and privacy. The decision of whether and how to offload tasks demands a basis in the investigation of these factors and the optimization of multiple performance metrics, including latency, throughput, energy consumption, and cost.

3. **Security:** Edge computing systems may face susceptibility to a range of security hazards, encompassing malware intrusions, data breaches, and denial-of-service (DoS) assaults. To fortify resilience, it is essential for edge computing systems to employ robust security measures, including encryption, access control, and anomaly detection and prevention. These measures serve to uphold credibility and safeguard service integrity. Moreover, the system should be intelligently designed to not merely identify well-known security threats but also to dynamically adapt to the evolving system landscape, thereby ensuring the prompt detection and prevention of novel forms of attack.

4. **Optimization:** Optimization is a key player in the composition of resilience for edge computing systems. It characterizes the technique of maximizing the effectiveness, performance, and scalability in edge computing systems while minimizing resource consumption, latency, and cost. Optimization is essential for ensuring that edge computing systems can meet the growing demand for real-time data handling, analytics, and decision-making in various applications, including IoT, smart cities, autonomous vehicles, and Industry 4.0.

   Optimization in edge computing involves several key factors, including resource allocation, task scheduling, load balancing, data placement, and network bandwidth management. By optimizing these factors, edge computing systems can improve their performance, reduce latency, and minimize resource consumption, thereby improving their overall efficiency and scalability.

   - Resource allocation encompasses assigning computing, storage, and networking resources to different edge devices and applications based on their requirements and priorities. By dynamically adjusting resource allocation, edge computing systems can optimize their performance, reduce energy consumption, and avoid overloading or under-utilization of resources.
   - Task scheduling involves scheduling different computational tasks to different edge devices based on their capabilities and availability, optimizing resource utilization, and minimizing latency. Load balancing covers the distribution of computational tasks evenly over several edge devices to prevent overloading and ensure fault tolerance.
   - Data placement involves placing data closer to the edge devices that need it, minimizing data transfer and latency, and reducing network bandwidth consumption. Network bandwidth management involves optimizing the use of available network bandwidth, prioritizing critical data, and avoiding congestion or network failures.

   Additionally, optimization in edge computing can be achieved using various approaches, including machine learning, deep learning, reinforcement learning, and metaheuristics. By using these techniques, edge computing systems can learn and adapt to changing conditions, predict future demands, and optimize their performance and efficiency in real time.

5. **Scalability:** Edge computing systems should be able to scale up or down to handle varying workloads and demands, especially during peak usage times or in the event of unexpected surges. In providing a resilient edge computing system,

the goal of scalability is to develop a system that has the capability to dynamically scale resources according to the fluctuations of incoming streams to achieve a minimal overhead and latency for the computing task.

6. **Heterogeneity and diversity:** Deploying edge computing systems across multiple locations or using multiple service providers can help ensure continuity and reduce the impact of localized disruptions. The distributed structure of edge computing systems helps prevent key failure spots, as in the case of cloud computing architecture. The failure of a node or group of nodes belonging to a particular cluster does not severely affect the overall performance of the system. Also, the system design should accommodate the system heterogeneity of the different IoT devices in a network. In fact, rather than poising issues of compatibility, the computing system should explore the devices' heterogeneity and maximize individual and collective performance.

7. **Resource management:** Edge computing systems should be able to dynamically manage the available resources, such as computation, storage, and bandwidth, to ensure that the system can adapt to changing conditions and workload. Computation-based resource management can optimize the performance of resource-constraint devices through computational offloading to minimize the computation on a device. It also addresses the task allocation strategies for energy-aware applications to provide resilience for devices with lower energy capacity. Efficient resource allocation schemes should be designed and managed and direct network traffics to avoid delay, which is crucial for latency-aware application. This can be achieved through resource allocation algorithms that optimize the utilization of resources while maintaining system resilience.

8. **Data management, recovery, and backup:** Edge computing systems should be able to efficiently store and manage data, especially in situations with limited or intermittent network connectivity. This can be achieved through data replication and synchronization techniques which affirm that data is always available and recent. The design requirement of good data management includes efficient data collection, data preprocessing, and data storage. The data management system should be integrated with other critical areas such as security, offloading, and disaster management and recovery. Furthermore, edge computing systems should have backup and recovery plans in place to quickly restore services in the event of a disruption or failure.

9. **Tracking and administration:** Continuous tracking and management of edge computing systems can support identifying potential issues before they cause disruptions and enable quick remediation.

In summary, **ECR** is a critical aspect of edge computing systems, especially in applications that require high availability, computational intensive, low latency, and real-time processing. By addressing the challenges mentioned above, edge computing systems can ensure that they remain operational and provide reliable services in the face of disruptions or failures.

Chapter 2
Scalability and Fault Tolerance for Real-Time Edge Applications

Abstract The rising demand of real time applications like health care, video surveillance and social media trend analysis, along with advancement in IoT and edge computing, ergs to use the edge computing for these real time applications. It can result in latency reduction, saving network bandwidth and efficient utilization of computing resources available at edge of the network. The nature of edge computing is distributed by-default hence it can offer a bench of resources for the applications executing in the edge computing, making it easy to handle the scalability requirements. Fault tolerance is of great concern for the real time applications in general and making it more important while working with edge computing. As, failures can result in disrupt of a critical activity, that can be more dangerous for a user or the real time application.

The devices in the edge network of an IoT system are usually deployed individually and normally separate from each other having different capabilities and requirements. The lack of redundancy plan during device deployment makes fault tolerance and scaling operations more important for the edge-computing environment. This chapter discusses in detail the related issues, including the causes of faults and strategies for mitigating faults by providing fault tolerance and scaling in the distributed edge environment.

Keywords Real-time applications · IoT integration · Latency reduction · Fault tolerance · Scalability · Bandwidth efficiency · Fault mitigation

2.1 Definition and Role of Fault Tolerance in Edge Computing

There are many big data applications, which process real-time data while ensuring low latency and providing the users with the fresh results. Including these applications are smart health care for patients, security surveillance, and social media analysis like Twitter and streaming social media. Edge computing offers a handy way to process these real-time applications efficiently. The edge nodes, which are typically distributed and offer bench of processing resources can help to meet the requirements of the big data applications. These heterogeneous and distributed edge nodes are more porn to failures requiring an efficient system to provide fault tolerance for the applications executing distributed in the edge computing.

Fault tolerance in edge computing means the IoT application executing in the edge-computing environment should be able to provide service in case of failure or any smart device acting as an edge or a typical server acting as an edge or a part of distributed edge network. Both real-time data streams as well as the processing tasks should be working smooth on other backup nodes or some replicated edge node to take over the control while any edge node goes down. There should be a framework, a mechanism, or some application that acts as a fault-tolerant agent that handles the smooth switching in case of any failure. Contemporarily this process should be automated and having most priority than any other task in the execution environment. The system should also ensure application timeline requirements as the real-time applications are time sensitive with respect to meeting their deadlines.

Tasks execution in edge computing is prone to all the faults related to distributed systems, during task arrival, scheduling and execution [1]. Moreover, the distributed edge computing environment executing a resource intensive task may face faults due to resource limitation of smart devices, mobility, heterogeneity and power issues. The services offloading [2], services duplication [3] and data replication [4] are used for failure handling of the individual task executing on an edge device. However, when we execute an application in distributed on edge network, the lack of availability of fault-tolerant systems for edge computing is a big issue. In addition, the failure rate for the devices in a distributed edge network is higher contrary to the traditional distributed systems. For resource limited edge devices, there is a great need for efficient fault tolerance system, which can provide reliability based on properties of edge devices, and in a decentralized manner, which is a key property of edge computing environment. Reliability and availability modeling and analysis are important requirements for data and processing to ensure robust design and operation.

The reliability at the edge network is concerned with successful delivery of services to the end user, and ensuring the latency requirements of IoT applications. If the reliability of tasks executing in edge computing environment is low then the user frequently experience service degradation or system failure. On the other hand, if the reliability is high then the resource-limited edge devices experience long service time this will introduces application delay. Therefore, both reliability and latency requirements are important factors to consider while designing a fault tolerance system for edge computing environment.

2.2 Causes and Types of Faults

The edge computing environment is advancing very fast and being used by many IoT applications to facilitate end users with quick results while keeping QoS. However, the hosted applications face service reliability and availability issues for both services providers and the end users. The edge computing is highly dynamic, distributed and decentralized in nature. Moreover, the heterogeneity of edge devices can result in several type of errors and failures in the edge network leading to performance degradation. Some renowned faults are inherited to the edge computing like following.

1. **Device faults:** These include node failure, limited services due to low power, hardware failures like CPU, memory, storage, sensor/actuator failure and communication port failure. Device out of range and no hardware support are also included in device faults.
2. **Network faults:** These include link failure, network partition, network congestion, communication failure, network timeout, packet loss, out of range errors (because of device mobility).
3. **Service faults:** These type of faults may occur, because of unavailability of specific service on the edge node, other possible reasons are resource shortage, software errors, services deadlock due to some required physical activity from user or some external world activity.
4. **Other faults:** Device/service migration or offloading not supported, environment hazards, sensor/actuators mismatch.

Failure in any system leads to services breakdown or complete shutdown. However, in edge computing there can exist partial failures due to individual device or a specific service failure, and for this designing a centralized system is not vise. Though these faults can still effect the overall system but a decentralized fault tolerance mechanism can increase the system efficiency [5]. Fault tolerance keeps the system functioning and provide uninterrupted services to the end users, and a fault-tolerant system is related to ensure reliability, availability, and avoidance of services breakdowns.

One should also consider features of an IoT application while thinking about fault tolerance for the edge computing, because, the edge nodes are executing tasks mostly belong to an IoT application. Here some key features of an IoT application are provided.

- **Real-time interaction:** Various IoT applications require real-time interaction with devices of some physical interaction of humans to complete specific tasks, such as health care, real-time traffic monitoring and security surveillances systems.
- **Low latency:** IoT applications require low latency communication and processing, the timely delivery of results is required to increase the quality of service (QoS) and to fulfill necessary requirements.
- **Geographical distribution:** Some IoT applications have distributed deployments to deliver services to both mobile and stationary users.
- **Support for mobility:** Mobility support is vital for lot of IoT applications to enable direct communication with mobile devices as well as to serve users while traveling.

- **Location awareness:** Location awareness is required to identify different objects and their position to fulfill critical requirements, for example object identification and tracking for security surveillances.
- **Fault-tolerant:** The IoT applications are normally required to interact with smart devices having low resources or mobility property attached to them, this can result in failure or unavailability of these devices. An efficient fault tolerance system is required to ensure the availability of an IoT application.
- **Big tasks:** Because the end devices are generating continuous streams of data so some tasks of IoT applications require high end processing.
- **High communication required:** Some IoT application require exchanging high data transfer rates like applications involving video streams and analysis, or social networking applications.
- **Working in heterogeneous devices:** An IoT application can involve a huge number of devices from different manufacturers with heterogeneous resources in terms of hardware and software.

2.3 Scaling for Distributed Edge Systems

The execution environment of the edge computing to process real-time applications that usually consists of streaming data tries to execute application tasks locally near to the user space before processing at cloud, resulting in decreased network overhead consequences, application delay, data security and privacy matters. Comparing the plentiful of resources at the cloud processing nodes in the edge network are low power devices and heterogeneity assassinated with them, along with device mobility. Executing resource intensive IoT applications on individual edge nodes can hamper the quality of service (QoS) and user satisfaction. However, distributed execution in edge computing can successfully execute a resource demanding, real-time, latency-oriented task by dividing its workload among available devices. This will advantage the application from the edge computing environment as well.

To leverage computing resources of edge nodes in an edge computing environment, mobile edge clouds [6], mobile device cloud [7], and Foglets [8] are proposed to coordinate numerous edge devices for resource intensive applications that are difficult to execute successfully on single edge device. Firework [9] leverages mobile devices and the cloud to process big tasks, it also combines different edge nodes to accomplished big data processing tasks cooperatively.

The stream processing big data applications including credit card monitoring systems, Global Positioning System (GPS) based monitoring systems, video surveillance systems, and online social network analysis usually process the incoming streams of data using distributed processing frameworks in a distributed computing environment. The stream processing systems (SPS) including Storm [10], Apache S4 [11] and Apache Flink [12], supports complex analysis of streaming data by using continuous queries to provide results in a very short span of time. These queries consists of several operators including stateless (filter) and stateful (aggregation, join) operators.

Input data streams are normally subject to fluctuation in data arrival rates and uneven workload variations causing some operators to be overloaded and finally result as a bottleneck operator, which can cause a failure in the system. There must exist and efficient mechanism for scale out and fault tolerance in stream processing application, especially when queries contain stateful operators, because stateful operators requires state migration to preserve the consistency of the operations [13]. For a cost effective scale out operation, it requires to identify legitimate bottleneck operator at first and then perform a scaling operation. The bottleneck detection is also important for fault tolerance as the bottleneck operator can cause error or failure for the stream processing application.

A horizontal scaling operation can be performed but most of current SPS consider either operator are stateless or put burden on application developer to handle state. Some research [14, 15] proposed methodologies for scaling a stateful operators but their solutions are based on static configurations or heuristics based assumptions which may result an improper scale operation, some other [16] also introduced to wait until some stateful operator fails due to overload and then perform failure recovery and scale out at same time.

2.4 Bottleneck Detection and Scaling for Fault Tolerance

Reason behind the scale out operations in SPS are bottleneck operators as these operators increase overall latency in applications and do not allow SPS to increase throughput. Most of the systems focus on how to scale out different type of operators based on an on-demand infrastructure or use some static measures of resources consumption of some use heuristics to identify bottleneck operator [16]. This may result in invalid detection leading to wrong scale out operation.

At first, it is required to find out the right bottleneck operator at runtime using proper measures of CPU utilization, memory usage and data size and then scale out only this operator. Wu Xu in [17] proposed a fuzzy logic based runtime bottleneck inspection, by using his technique we can place some check to those operators, which are going to bottleneck. By using the fuzzy logic base system to run time bottleneck inspection and during the process of defuzzification we calculated the center of gravity of output signal obtained after the evaluation of the different rules ranging from 0 to 100. This range is divided in three sections corresponding to scale by imposing two thresholds (δ_1, δ_2), i.e. a node can be chosen for scale out if $O(X) > \delta_2$ ($scale\_out\_threshold$), or scale in when $O(X) < \delta_1$ ($scale\_in\_threshold$) and keep unchanged in other cases [17].

We have targeted δ_2 and added another parameter the scale from 0 to 100 and called it $alarming\_threshold$ (δ_2'). We assume that operators which are at δ_2' can be bottleneck operators in near future, and at this stage we start backing up state of this operator as depicted in Fig. 2.1.

To test a node is overloaded or not Algorithm 1 is presented which calculates the $alarming\_threshold$. Here if statement checks for $Scale\_out\_threshold$ to

Fig. 2.1 Operator load

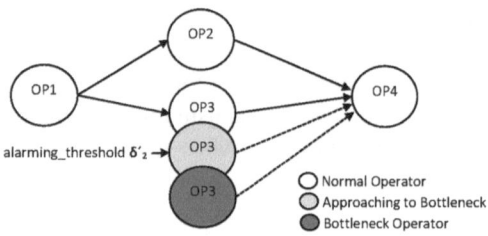

test if node is bottleneck in line 14, then activate SE setup as primary. In addition, on line 17, if statement checks that either the node is approaching bottleneck or not in near future by checking *alarming_threshold* and if it happens, it starts SE as active backup server. It should also be considered that wastage of resources have to be avoided if current resources are working fine. Hence, we added another check on node to test if SE was started in the past on approaching node at *alarming_threshold*, and now node is working as normal routine then stop SE as active backup; this is performed in line 19 and 20. Line 21 checks for when the system has to perform scale in by tracking *Scale_in_threshold*. Details of bottleneck detection algorithm can be found in our previous article [17].

2.4.1 Essence of Bottleneck Detection

Scaling Distributed Stream Processing Systems (DSPS) while considering state management for stateful operators have recently gained much interest in both academia and industry [15, 16, 18, 20]. Scaling out stateless operators can be performed by starting a new operator instance with a blank "memory", but scaling operation of stateful operators requires state migration to preserve the consistency of the operations [13]. One of the biggest challenges is to define a proper method to identify the bottleneck operator and then scale it out by offering lowest latency. Some work used measure of CPU utilization of processing nodes as threshold elasticity policy [16, 19, 20], but only one measure cannot prove that operator is bottleneck operator as CPU utilization can increase or decrease very frequently. Stela [20] presented an elasticity technique for stream processing systems with creating a novel metric, ETP (Effective Throughput Percentage), that accurately captures the importance of operators based on congestion and contribution to overall throughput. We have used their mechanism along with fuzzy logic system used by Raphael Frank et al. [21] to detect bottleneck operator [17].

Scaling with nearly zero latency is not a problem for stateless operators but for stateful operators it is very difficult because it should be application transparent and with minimal overhead. Gulisano et al. [22] describes how to partition state and decide when to scale out by monitoring incoming load and resource utilization but migrating a stateful operator requires synchronization in state recreation protocol making a complex policy at overall.

Algorithm 1 Bottleneck detection using fuzzy logic

Require: Operator O_i (where i= 1, 2, ..., k) to process incoming streams;
    FIS as fuzzy logic engine;
    Status as status of nodes;
Ensure: Detect bottleneck operator;
    Parameters   for   the   operator   *Alarming_threshold*,   *Scale_out_threshold*,
    Scale_in_threshold;
 1: **for each** node **do**
 2:     init FIS with $jFuzzyLogic$ library
 3:     FIS.load (fuzzy_rules)   ▷ FIS load fuzzy rules from file
 4:     **if** init FIS = fail **then**
 5:         throw Exception
 6:     **else**
 7:         FIS.getFunctionBlock()
 8:         **while** node is running **do**
 9:             **if** !(get.Status(node)) **then**
10:                 throw Exception
11:             **end if**
12:             FIS.setVariable(Status)
13:             $Result \leftarrow$ FIS.evaluate()
14:             **if** $Result \geq Scale\_out\_threshold$ **then**
15:                 Stop this node
16:                 Start SE as Primary
17:             **else if** $Result \geq Alarming\_threshold$ **then**
18:                 Start SE as active backup
19:             **else if** $(Result < Alarming\_threshold)\&\&(Result > Scale\_in\_threshold)$
        then
20:                 Stop SE as active backup
21:             **else if** $Result \leq Scale\_in\_threshold$ **then**
22:                 This node should scale in
23:             **else**
24:                 keep running
25:             **end if**
26:         **end while**
27:     **end if**
28: **end for**

2.5   Scale Out Methods

Scalability of stateful operator, i.e. the ability to dynamically scale resources according to the fluctuations of incoming streams with minimal overhead and latency. Fault tolerance and scalability both are essentially required for replicating or checkpointing the computation state of the operators among different nodes. At any point, failed operators create bottleneck and can be recovered by using our proposed techniques as per requirements. The active backup approach will help us to achieve almost zero latency goal while the check pointing based technique is resource efficient technique.

For the scaling methodology horizontal scaling methodology is handy to adopt due to the nature of the edge computing environment. The important thing here is to define some modeling for the incoming data streams and their respective operators, which process the real-time application logic. Moreover the state model for these operators also needs special attention.

2.5.1   Stream Model and Operator Model

Data stream is an infinite sequence of tuples $t(t \in S)$ which is denoted as $S = (t_1, t_2, \ldots, t_n)$ and each tuple has following schema $t = (\tau, \kappa, \upsilon)$ here τ denotes logical time stamp of the tuple at which it is received and κ is key field and υ denotes value of the tuple. Figure 2.2 illustrates our stream model.

A query Q which consists of a network of operators is defined by Directed Acyclic Graph (DAG) denoted as $Q = (O, S)$ where O is set of operators and S is stream of tuples flowing between operators. Tuples are processed by operators that belong from some continuous stream S. An operator O takes n input streams denoted by set $I = \{s_1, s_2, \ldots, s_n\}$ process their tuples and produce one or more output streams. For ease of presentation we restrict to a single output stream. Working of an operator O can be defined using the function $f(I, \tau, \theta) = (O, \tau, \theta)$.

The function f accepts finite set of input streams I process their tuples up to time stamp τ and produces output stream O until timestamp τ. Here θ denotes state of the operator if operator is a stateful operator, in case of stateless operator value of θ will be null. Figure 2.3 illustrates our model for both stateless and stateful operator.

2.5.2   State Model

Stateful operators are normally aggregation operators or join operators and in both of these operators operations are performed based on some key and values of these keys manipulated accordingly. So we can assume that these keys can be unique for these type of operators. This concept will help us to maintain state of the operator during processing and migration. We define state θ_o of an operator as set of key/value pairs i.e. $\theta_o = \{(k_1, v_1), \ldots\}$. Figure 2.4 illustrates concept of key/value state of an operator. Each key is unique as well as independent of other keys and refers to corresponding tuple key from the input streams, this will also

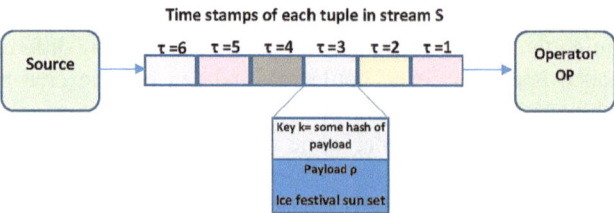

Fig. 2.2 Stream model

Fig. 2.3 Operator model

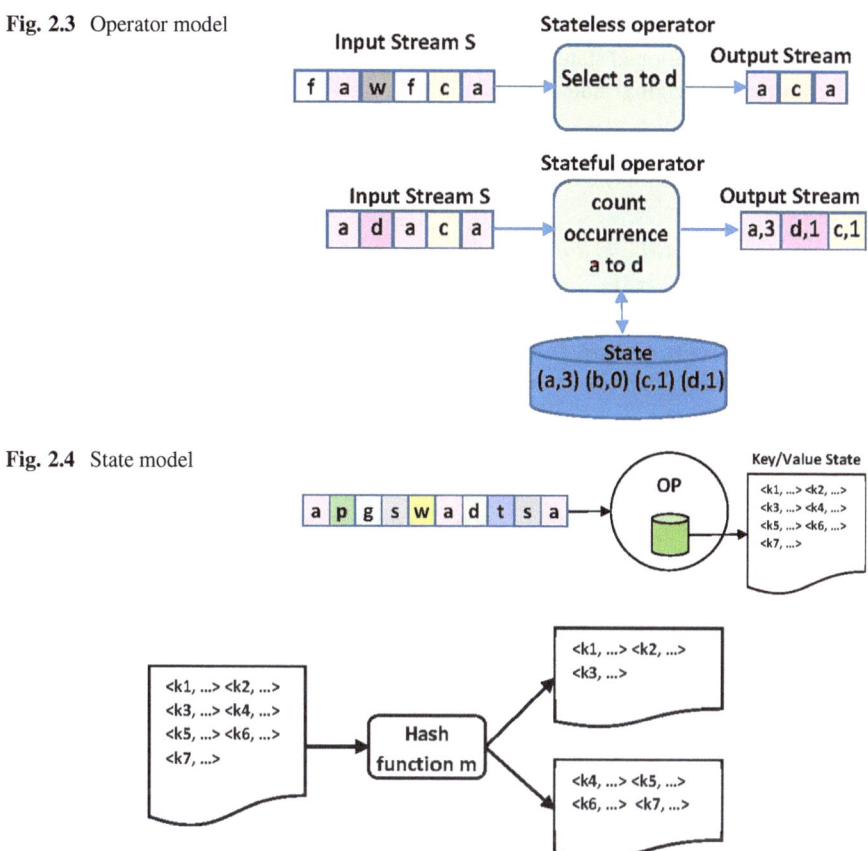

Fig. 2.4 State model

Fig. 2.5 State partition model

lead to independent state of each operator. Its associated value v stores the portion of processing state for that operator. While partitioning an operator to η new nodes we also have to partition their key space to these new nodes so that each partitioned operator will be working independently from others. We can partition this key space using hash function $m : k \rightarrow [1, \eta]$. Figure 2.5 shows partitioned states.

During execution at time t, the stateful operator OP have some memory $M = (\theta, \kappa_{range}, S)$ where κ_{range} is key range of OP and S is size of memory. During scale out we have to decompose $M = (\theta, \kappa_{range}, S)$ to p partitions where $p \in \mathbb{Z}$. When M is decomposed the resulting new partitions will also become state for newly partitioned operators. Furthermore, the composition of these partitions yields state of OP that is semantically equivalent to original. For two partitions decomposition is shown in Eq. (2.1)

$$compose(M_0', M_1') = M \leftrightarrow Decompose(M, p) = (M_0', M_1') \qquad (2.1)$$

We say partitions are semantically correct if read operation for some key κ is performed on partitioned state the resulting output should be same if read operations is performed on original state.

$$Let\, M' = Decompose(M, p)$$

$$read_{decompose}(M'_0, M'_1, \kappa) \equiv read(compose(M'_0, M'_1), \kappa) \equiv read(M, \kappa)$$

2.5.3   Scale Out Model

To scale out a bottleneck operator OP_i, we will split it into η new nodes based on required degree of parallelism to resolve bottleneck. It is given by the following:

$$OP_i = OP_{i1}, OP_{i2}, ..OP_{i\eta} \tag{2.2}$$

We will also partition state of the operator and its input streams. While partitioning input streams, we will add sequence number so that we can reorder these streams before forwarding these streams to next operator, as depicted in Fig. 2.6.

To handle scale out for any operator OP_i, a fair scale out means that resource allocation should be done well in time to solve the bottleneck.

For OP_i at time t, current assigned resources $\mathcal{R}cur_{op}(t_k) \in [0, t_k]$ are consumed, which can be cumulatively shown in Eq. (2.3).

$$\mathcal{R}cur_{op}(t_k) = \int_0^{t_k} cur_{op}(t)dt \tag{2.3}$$

where $cur_{op}(t)$ are the resources currently consumed by OP_i at time t. To resolve the bottleneck, total number of needed resources $\mathcal{R}need_{op}(t_k) \in [0, t_k]$ can be

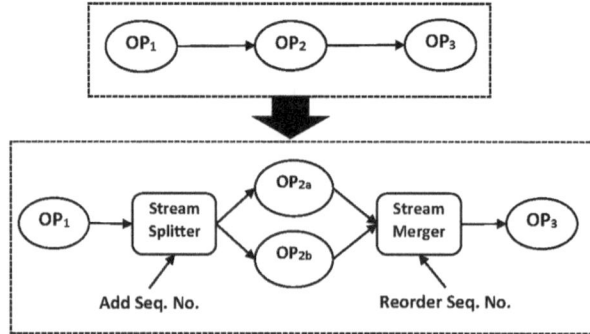

Fig. 2.6 Operator scale out model

cumulatively defined as Eq. (2.4).

$$\mathcal{R}need_{op}(t_k) = \int_0^{t_k} need_{op}(t)dt \tag{2.4}$$

where $need_{op}(t)$ are total required resources by OP_i at time t. The fairness degree fd_{op} for operator OP_i at time t_k is given by Eq. (2.5).

$$fd_{op}(t_k) = \frac{\mathcal{R}cur_{op}(t_k)}{\mathcal{R}need_{op}(t_k)} = \frac{\int_0^{t_k} cur_{op}(t)dt}{\int_0^{t_k} need_{op}(t)dt} \tag{2.5}$$

To resolve a bottleneck, total needed resources $\mathcal{R}$ $need_{op}(t_k)$ are always greater than currently assigned resources $\mathcal{R}$ $cur_{op}(t_k)$, hence the fairness degree $fd_{op}(t_k)$ $\in [0, 1]$ holds. It shows that if $fd_{op}(t_k) = 1$, then at time t_k, all needed resources are provided, while a 0 value shows that the operator is still bottleneck because needed resources are not provided yet.

To obtain average fairness degree for an application (i.e. $Fd_{app}(t_k)$) in which n operators are bottleneck at time t_k, is given by Eq. (2.6).

$$Fd_{app}(t_k) = \frac{1}{n}\sum_{n=1}^{n} fd_{opi}(t_k) \qquad where \qquad fd_{op}(t_k) \in [0, 1] \tag{2.6}$$

2.6 State Backup and Scale Out

State backup can be performed using active standby, passive standby, and upstream backup. In the active standby approach, both primary and secondary nodes are given same inputs and two nodes are processed similarly, meaning state and outputs are also the same. Output of primary node is connected to downstream node, while output of secondary node is not connected. In case of secondary takeover, it requires only connecting its output to downstream node, and this has almost nearly equal to zero delay but requires extra resources, and the cost is same as that of a primary server. In the passive standby approach, the backup server is always behind the primary server as the primary task saves its state periodically (checkpoint) to a permanent shared storage and this information is used to create state of secondary server when it takes over. So passive standby or checkpointing may offer some delay in overall system performance. The next section discusses how we have handled these techniques efficiently and effectively.

2.6.1 Active Backup

Our active backup server technique is not as that of costly as that of we are making backup server for each primary server rather we are using this technique with slight changes. During bottleneck detection when an operator approaches at *alarming_threshold* (δ_2') scale out is performed by starting secondary servers using SE module. This module will be responsible for partitioning state and input streams of primary server to new nodes. The benefit of active backup technique that there is no need to store backup on upstream servers in the form of checkpoints which may result in overloading upstream server also. The other benefit is that there is no need to replay some tuples to achieve state similar to one running on primary server, which will result in time saving as well as delay avoidance. One problem of active backup technique is that it doubles the cost, but in our implementation the SE will not start any new nodes until we detect some operator approaching near to the bottleneck. So this means that we are not always running extra server as backup of some primary server rather we start extra server only when it is required.

In Fig. 2.7 execution plan is shown with addition of SE module. As we are continuously tracking load of the node so suppose workload of OP_3 is started to increase and at time t this operator approaches to *alarming_threshold* (δ_2') now SE will start scale out by adding two new nodes OP_{3a} and OP_{3b} and when OP_3 reach at *scale_out_threshold* SE will replace primary server with new nodes. This active backup is not working all the times (i.e. from start of OP_3) and we are starting this active backup when OP_3 will reach at (δ_2') this will help us to reduce infrastructure cost and at the same time by using SE we are aiming to achieve zero latency in case of operator bottleneck.

Regarding working of SE first we will copy state from primary node running OP_3 and partition it further on two new nodes as OP_{3a} and OP_{3b} so that we can divide load and process OP_3 with more speed (number of new nodes can be increased or decreased as per requirement). We also have to partition data stream S_3 into S_{3a} and S_{3b} accordingly so that both new operators should get their respective data streams for processing tuples. We should take care of splitting stream from tuple t_i with time τ_i which is the time when the state θ_o is copied to SE from OP_3.

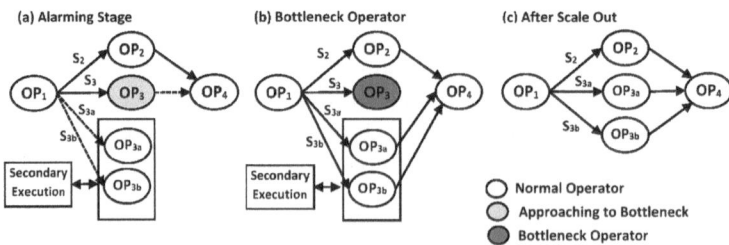

Fig. 2.7 Active backup and secondary execution

Algorithm 2 Active backup and scale out

Require: On approaching operator O at *alarming_threshold* Start SE;
    copy_state(θ) to SE;
    copy_timestamp(τ) to SE;
Ensure: Scalability using active backup
 1: **for each** input streams to O **do**
 2:    copy(tuple_ID&& stream_ID) to SE
 3: **end for**
 4: Partition key space to η new nodes in SE
 5: **for each** η_i in SE **do**
 6:    **while** there exist key k in key space **do**
 7:        **if** hash(k)=η_i **then**
 8:            $\eta_i \leftarrow k$
 9:        **end if**
10:    **end while**
11: **end for**
12: Partition copied state θ_{copy} to η new nodes in SE
13: **for each** η_i in SE **do**
14:    create new state θ
15:    **for each** key k assigned to η_i **do**
16:        $\theta \leftarrow$ assign (k, v) ▷ for all keys $k \in \theta_{copy}$
17:    **end for**
18: **end for**
19: Partition input stream S_o to S_η
20: **for each** tuple t in S_o **do**
21:    **for each** η_i in SE **do**
22:        **if** $k_t = k_{\eta_i}$ **then**
23:            route tuple t to η_i
24:        **end if**
25:    **end for**
26: **end for**

Algorithm 2 describes how state management is performed by first copying state and then by partitioning key space, the state itself and incoming streams to new nodes in SE. During initialization when an operator reaches *alarming_threshold* SE is started and copies state and time stamp τ to SE. As a single operator can process multiple input streams so there is a need to copy *stream_ID* and *tuple_ID* of respective streams to SE this is performed using a loop in line 1 to 3. After copying now we have to divide key space from line 4 to line 11 and this is done by using a hash function to partition key space to η new nodes, all keys are partitioned evenly to new nodes in SE setup so that workload is divided evenly among all new nodes. In line 14 a new state θ is created on each new node and we partitioned copied state θ_{copy} to η new nodes and assigned respected key value pairs (k,v) to newly created state θ (this is performed in line 15 and 16). Finally we have partitioned input streams of operator to available nodes in SE from line 20 to 26, we have to be careful here that from incoming streams all tuples t should be routed to that η_i node which have relevant partition of key space.

2.6.2   Checkpointing and Scale Out

Checkpointing saves state of operator to some storage which allows it to be further processed in the form of scaling out an operator or restarting some failed operator from that stored state, this process is known as checkpointing. Raul Castro et al. [16] performed scale out and fault tolerance in stream processing using operator state management, it says that operator state can be check pointed periodically by the SPS and backed up to upstream servers. After bottleneck of some operator this backup can be used to scale out or in case of failure recovery can be performed. Problem with this approach is that system is performing continuous checkpointing for every operator which introduces an extra overhead on each operator. Also, the backup is stored on some upstream operator, which puts extra load on upstream operators. To handle this problem we are proposing a solution where continuous checkpointing is not allowed for every operator; rather, we are checkpointing only those operators which are approaching to *alarming_threshold* (δ_2') stage, and also not storing backup on upstream server. We have introduced a SM module for the checkpointing and scale out technique, for scaling the bottleneck operator. SM will handle the state first by storing its backup, which will be obtained by checkpointing the state and afterwards partitioning this state to new nodes. The checkpointed state will contain (θ_o, τ_o, β_o) where θ_o is state of operator, τ_o is time stamp of most recent tuple which is processed and reflected in state and β_o is buffer attached to the operator to temporarily hold tuples flowing from upstream operator to downstream operator. Remember the system will not always be doing checkpointing. The process of checkpointing will start when the operator approaches to *alarming_threshold* (δ_2'). The checkpoint state is executed now onwards asynchronously and triggered at every check pointing interval c, or after a user defined event, for example, when the state has changed significantly. After the operator state was backed up, the processed tuples from output buffers in upstream operators can be discarded because they are no longer required to process.

Suppose that in Fig. 2.8 OP$_3$ is near bottleneck, which means it is at *alarming_threshold* (δ_2'), we have to start check pointing for this operator to SM, and when this operator reaches at *scale_out_threshold* (δ_2), SM will start new nodes

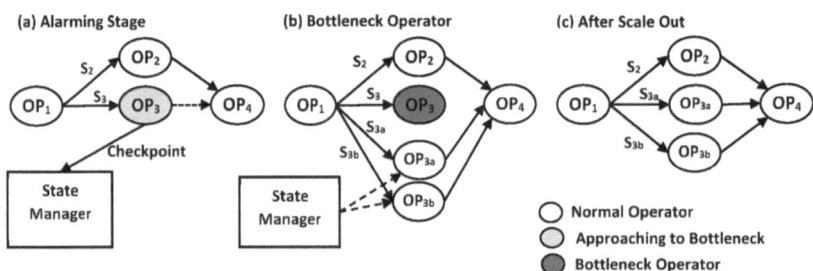

Fig. 2.8 Checkpointing and scale out

OP_{3a} and OP_{3b} and partition the checkpointed state and tuples stored in β_o to new nodes and process these tuples. At the same time, also partition input streams and route the partitioned streams to new nodes and starts processing new nodes.

Algorithm 3 Checkpointing and scale out

Require: On approaching operator O at *alarming_threshold*;
    checkpoint $(\theta_o, \tau_o, \beta_o)$;
    SM Partition key space to η new nodes;
Ensure: Failure recovery or scalability using checkpointing
 1: Partition key space to η new nodes in SE
 2: **for each** η_i in SE **do**
 3:    **while** there exist key k in key space **do**
 4:        **if** hash$(k)=\eta_i$ **then**
 5:            $\eta_i \leftarrow k$
 6:        **end if**
 7:    **end while**
 8: **end for**
 9: Partition θ_o to η new nodes
10: **for each** η_i in SE **do**
11:    create new state θ
12:    **for each** key k assigned to η_i **do**
13:        $\theta \leftarrow$ assign (k, v) ▷ for all keys $k \in \theta_{copy}$
14:    **end for**
15: **end for**
16: Partition tuples in buffer β_o to η new nodes and process them
17: **for each** tuple t in β_o **do**
18:    **for each** η_i in SE **do**
19:        **if** $k_t = k_{\eta_i}$ **then**
20:            process tuple t on η_i
21:        **end if**
22:    **end for**
23: **end for**
24: Partition input stream S_o to S_η
25: **for each** tuple t in S_o **do**
26:    **for each** η_i in SE **do**
27:        **if** $k_t = k_{\eta_i}$ **then**
28:            route tuple t to η_i
29:        **end if**
30:    **end for**
31: **end for**

Algorithm 3 describes how the process of checkpointing and scale out works. On approaching some OP at *alarming_threshold* (δ_2') we start backing up its state using checkpointed state $(\theta_0, \tau_0, \beta_0)$ and store it to SM. When operator O approaches to *scale_out_threshold* (δ_2) the state manager will start partitioning its key space to η new nodes by using hashing mechanism (line 1 to 8). Now SM will partition state of operator by obtaining a copy from most recent checkpointed state, for this purpose SM first creates an empty state on each new node (line 11) and then assign key value pairs from θ_0 to newly created state (line 13) this will complete state partitioning process. Now we will distribute tuples stored in buffers β_o towards respected node (line 19) and process these tuples on that node (line 20). Now new nodes are ready to process input streams, but we will partition the input stream S_o to S_η (line 24 to 31). New nodes have started processing input streams for which they will produce some output streams, we have to make sure when output streams of new nodes and output streams of the bottleneck node are same, once this achieved we will stop bottleneck node and route output streams of new nodes to downstream operator.

2.7 Integration Design

We have added our state management system using Apache Storm which is a poplar stream processing system. It is free and open source project under umbrella of Apache Software Foundation.

2.7.1 Apache Storm

Some key abstractions in Apache Storm includes tuple which is ordered list of key-value pairs and a stream which is an unbounded sequence of tuples. A streaming application is represented by a topology called Directed Acyclic Graph (DAG) that contains operators as nodes and links between operators denotes flow of streams. Operators can be spouts or bolts, which contains actual processing logic, spouts represents source of streams while bolts process input streams to produce results. Bolt can run functions like filter, aggregate, join or connectivity to some database.

There are two types of nodes in Storm Nimbus and Supervisor. Nimbus is master node and is responsible for scheduling tasks to worker nodes. Once scheduling is done the plan is communicated to worker nodes using ZooKeeper [23], which is a shared memory service for communication among distributes systems. Supervisor is a slave node having one or more worker processes and it assigns tasks to workers, each task may run as single or multiple executors. A topology is submitted to Nimbus which gathers tasks to be executed and distribute these tasks to available Supervisors. Storm parallelizes operators to increase the throughput.

Bolts are by default stateless to handle stateful operators it requires maintaining state in the memory as Map or any other data structure, but when worker or node fails this state is lost. Currently state persistence is only achieved through regular checkpointing to a remote Redis [24] store. This feature was introduced in Storm 1.0.0 which has its own limitations described in Sect. 2.4.

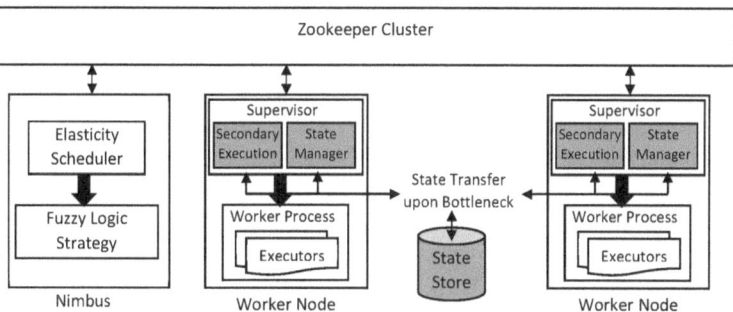

Fig. 2.9 Integration with Storm, where newly added modules are highlighted in grey

2.7.2 Integration with Storm

Our proposed dynamic state management framework is a lightweight module integrated with Storm to check for bottleneck operator and then scale it out by handling state using our proposed approaches i.e. active backup and checkpointing and scale out. Currently, Storm has 4 kinds of built-in schedulers and also allows to implement your own scheduler to assign executors to workers. We run a customized scheduler by configuring the class to use the "storm.scheduler" config in storm.yaml and also implemented IScheduler interface in Storm default scheduler, hence replacing default scheduler of Storm.

We used scheduler and fuzzy logic strategy modules from our previous work by Wu Xu [17]. The SE module is responsible for scaling operations related to our active backup technique, while SM module will handle scaling operations for checkpointing and scale out technique. We have also added modules for state storage and retrieval where we have used in-memory data store Redis [24]. Figure 2.9 shows architecture of the system integration with Storm.

2.8 Evaluation

Next we evaluate our approach for dynamic state management for stateful operators in SPS. At first, we have tested effectiveness of our approach using an application having stateful operators. Moreover, we tested our approach for application latency and application performance by analyzing throughput and latency measures. We have tested by enabling and disabling our approach and also by comparing the state persistence mechanism offered by Storm.

2.8.1   Experimental Setup

We run the experiments using Storm v1.1.2, using a cluster of 3 nodes each configured with 3 worker nodes, so total 9 workers are running. Among three nodes one machine is with 6GB RAM and other two having 4GB RAM with Core i5 CPU @ 2.30 GHz processors. We run additionally Nimbus and Redis services on machine with 6GB of RAM.

In order to test for effectiveness of our approach we tested an application of frequent pattern detection, and for pattern generation, we have used an offline stream data producer, JSON-Data-Generator (JDG) [25]. This is an open source project, which can generate a real-time stream of JSON data. We have integrated it with our application to generate different patterns and a detector operator (stateful operator) maintains a counter for the appearances of these patterns.

2.8.2   Results

To test for elasticity, we have tested both of our techniques in same fashion which is described below. In order to test for elasticity we configured stateful bolt to increase number of executors to double in case of bottleneck, starting at 2 executors for spout and 5 executors for stateless bolt and 2 executors for stateful bolt. We started with simple workload 200 tuples per second and increased it continuously by almost doubling the rate after every 500 seconds. Figure 2.10 shows scale out for our active backup technique. It depicts that each time we increased load bottleneck, was detected and scale out operation is performed. After scaling it out the number of executors were increased with respect to scale out ratio set for stateful bolt. During first scale out operation, number of executors raised to 13, but when the SE (in case of Active Backup) or in case of checkpointing option our newly added executors taken over the bottlenecked operator number of executors became 11. This is because in our implementation we continue to run bottleneck operator

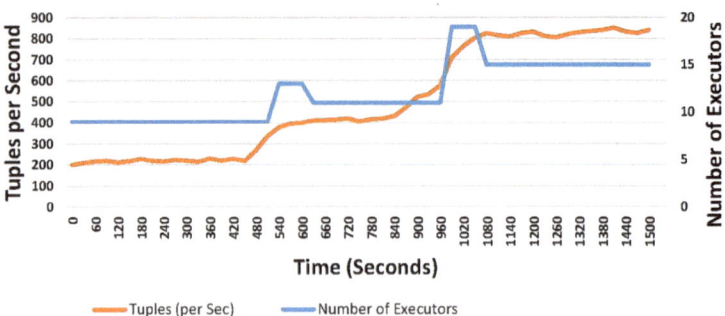

Fig. 2.10 Scale out for active backup

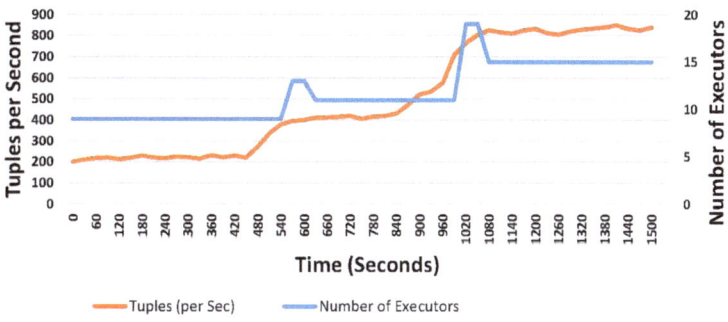

Fig. 2.11 Scale out for checkpointing

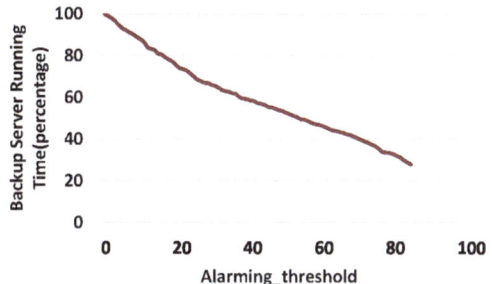

Fig. 2.12 Effect of $alarming\_threshold$

until backup has taken over. Once backup nodes started to work as normal nodes bottleneck operators were stopped to release the resources.

Figure 2.11 shows scale out for our checkpointing technique. Scale out behavior for both techniques is same, but there is only difference of resource consumption. Checkpointing techniques uses less resources as compared to active backup technique.

The fuzzy logic module is working in the system to find out the right bottleneck operator. Based on output of the system, we are using $alarming\_threshold$ (δ_2') and $scale\_out\_threshold$ (δ_2) to start out backup servers using SE module. We have tested $alarming\_threshold$ with respect to the percentage of time in which backup servers are running. Figure 2.12 shows that while using a lower value of $alarming\_threshold$, it will increase percentage running time of our backup servers and if we select a 0 value it shows our servers are running all the time, which results in wastage of resources. On the contrary, if we are increasing the value of $alarming\_threshold$ and it approaches a high value (i.e., near to $scale\_out\_threshold$), it will result in efficient usage of resources. It is not wise to use high value of $alarming\_threshold$ approaching to $scale\_out\_threshold$, because SE needs some time to set up backup servers, as the percentage of time for running backup servers is decreasing. The $scale\_out\_threshold$ is totally dependent on status of the operator detected by the fuzzy logic module.

We have tested latency and compared it with state checkpointing mechanism implemented by Storm 1.0.0. Keep in mind that Storm still does not offer to scale out automatically, so we have to restart topology with more resources to handle increased workload. This is clear in Fig. 2.13 that when input rate is increased, latency for Storm checkpointing is increased with very large rate until topology is restarted. On the contrary, when scale out is performed automatically using our techniques, the increased workload is handled efficiently.

The throughput graph in Fig. 2.14 shows that our techniques shows more throughput when compared with Storm's checkpointing mechanism, where the overall throughput remains low. It is also clear that during high input rate elasticity mechanism implemented using live migration technique shows better results among all other compared approaches. Around time 680 and 1050 seconds, throughput of Storm checkpointing approaches to zero due to restart of topology handle increased workload.

Both of these results related to application latency and throughput prevail that both of our techniques are showing almost same behavior, but live migration approach is somewhat in better position. This is due to the fact we are not starting state management tasks from start of application, but only when it is required. Results also show that latency in our checkpointing based scale out mechanism is greater. This is because this technique requires replaying some tuples stored in buffers. This has supported our assumption that live backup technique is better if low latency is required but it will result in little bit of more resources cost as we have to start live backup nodes before an operator is fully bottleneck. And if user want to have low resource cost and he can afford latency during scale out, then he can use checkpointing scale out mechanism.

To explore state management overhead, we run the application continuously for 30 minutes by alternating low input rates and high input rates after every 5

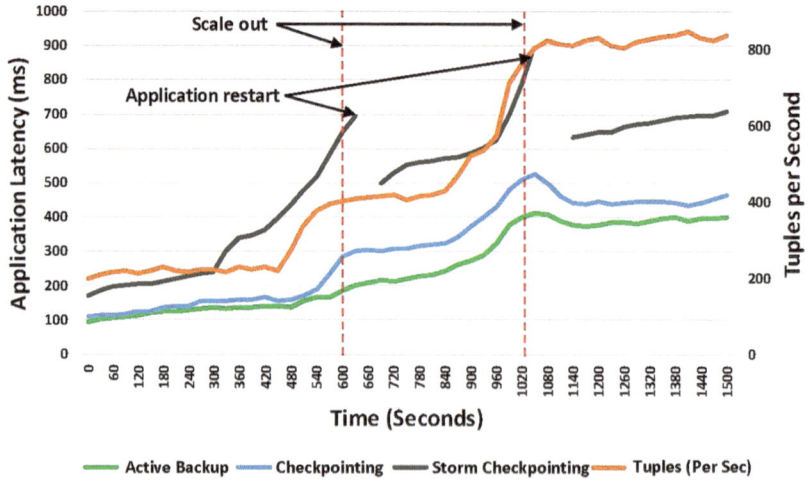

Fig. 2.13 Latency comparison

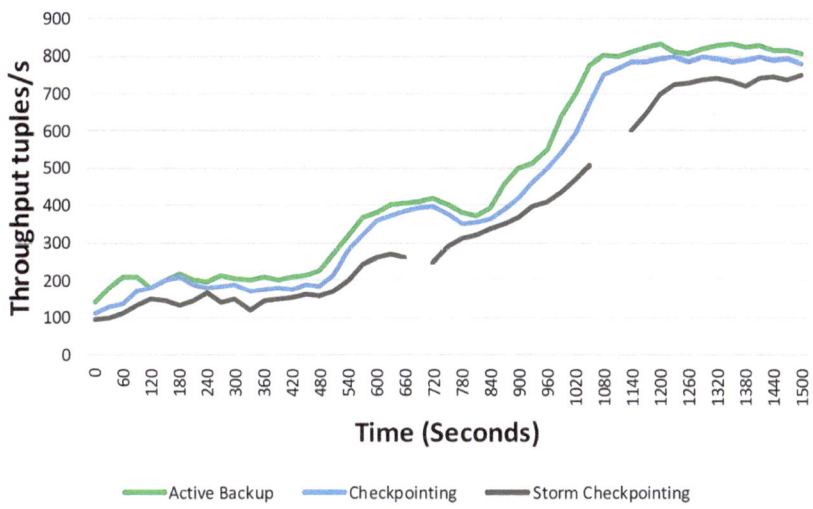

Fig. 2.14 Throughput comparison

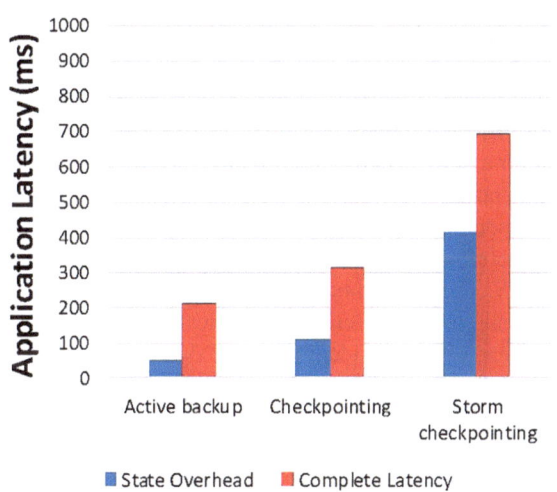

Fig. 2.15 State management overhead in first scale out

minutes, and calculated latency occurred due to state management. Figures 2.15 and 2.16 both show that our techniques are performing better as latency for them is low with respect to Storm's checkpointing mechanism; this is because that state checkpointing is perform after every second which introduces extra overhead and results in more latency. On the contrary, our techniques only store and retrieve state when required, that is, when and operator approaches to bottleneck.

Fig. 2.16 State management overhead in second scale out

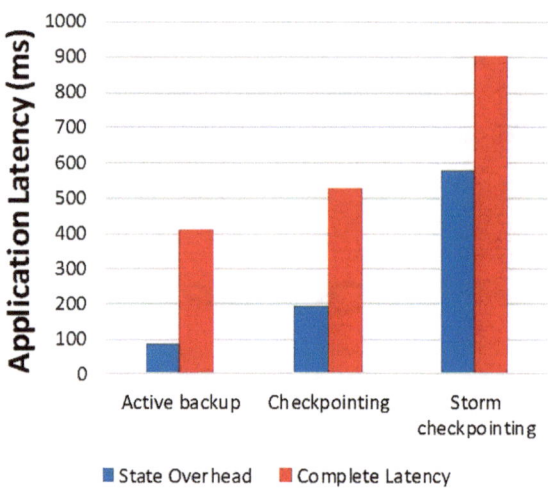

2.9 Summary

In this chapter, we have provided techniques to avoid a failure in the stateful operator while processing a latency aware streaming application. Two approaches were proposed and implemented named as active backup and checkpointing, to handle runtime scale out for a bottlenecked stateful operator. After detecting a bottlenecked operator correctly, the first technique enables SPS to achieve nearly zero latency during scale out operation by using active backup technique, while the second one targets resource efficiency. We have integrated our approaches into Storm, a popular parallel SPS. The experimental results shows effectiveness of our solution by providing live scale out and decreasing application latency by minimizing state management overhead. We also identified that overhead of state management is very low for both of our techniques as state management operations are performed only when required, compared to checkpointing mechanism proposed by Storm that performs state management continuously.

References

1. Bruning, S., Weissleder, S., Malek, M.: A fault taxonomy for service-oriented architecture [C]. In: IEEE High Assurance Systems Engineering Symposium (2007)
2. Flores, H., Hui, P., Nurmi, P., et al.: Evidence-aware mobile computational offloading [J]. IEEE Trans. Mob. Comput. **17**(8), 1834–1850 (2017)
3. Mohamed, N., Al-Jaroodi, J., Jawhar, I.: Towards fault tolerant fog computing for IoT-based smart city applications [C]. In: 2019 IEEE 9th Annual Computing and Communication Workshop and Conference (CCWC), pp. 0752–0757 (2019)
4. Qaim, W.B., Ozkasap, O.: DRAW: data replication for enhanced data availability in IoT-based sensor systems [C]. In: 2018 IEEE 16th Intl Conf on Dependable, Autonomic and

Secure Computing, 16th Intl Conf on Pervasive Intelligence and Computing, 4th Intl Conf on Big Data Intelligence and Computing and Cyber Science and Technology Congress (DASC/PiCom/DataCom/CyberSciTech), pp. 770–775 (2018)

5. Mudassar, M., Zhai, Y., Liao, L., Shen, J.: A decentralized latency-aware task allocation and group formation approach with fault tolerance for IoT applications. IEEE Access **8**, 49212–49223 (2020)

6. Fernando, N., Loke, S.W., Rahayu, W.: Computing with nearby mobile devices: A work sharing algorithm for mobile edge-clouds. IEEE Trans. Cloud Comput. **7**(2), 329–343 (2019)

7. Fahim, A., Mtibaa, A., K.A. Harras: Making the case for computational offloading in mobile device clouds. In: Proceedings of the 19th Annual International Conference on Mobile Computing and Networking, ser. MobiCom '13. Association for Computing Machinery, New York, NY, USA, pp. 203–205 (2013) [Online]. Available: https://doi.org/10.1145/2500423.2504576

8. Saurez, E., Hong, K., Lillethun, D., Ramachandran, U., Ottenwalder, B.: Incremental deployment and migration of geo-distributed situation awareness applications in the fog. In: Proceedings of the 10th ACM International Conference on Distributed and Event-Based Systems, ser. DEBS '16. Association for Computing Machinery, New York, NY, USA, pp. 258–269 (2016) [Online]. Available: https://doi.org/10.1145/2933267.2933317

9. Zhang, Q., Zhang, Q., Shi, W., Zhong, H.: Firework: Data processing and sharing for hybrid cloud-edge analytics. IEEE Trans. Parallel Distrib. Syst. **29**(9), 2004–2017 (2018)

10. Apache Storm [EB/OL]. http://storm.apache.org/Accessed:2020-04-02

11. Neumeyer, L., Robbins, B., Nair, A., et al.: S4: Distributed Stream Computing Platform [C]. In: 2010 IEEE International Conference on Data Mining Workshops, pp. 170–177 (2010)

12. Apache Flink [EB/OL]. https://flink.apache.org/Accessed:2020-04-02

13. Gedik, B., Schneider, S., Hirzel, M., et al.: Elastic scaling for data stream processing [J]. IEEE Trans. Parallel Distrib. Syst. **25**(6), 1447–1463 (2014)

14. Ding, J., Fu. T.Z.J., Ma, R.T.B., et al.: Optimal operator state migration for elastic data stream processing [J]. Hal Inria **22**(3), 1–8 (2015)

15. Cardellini, V., Nardelli, M., Luzi, D.: Elastic stateful stream processing in storm [C]. In: 2016 International Conference on High Performance Computing Simulation (HPCS), pp. 583–590 (2016)

16. Castro Fernandez, R., Migliavacca, M., Kalyvianaki, E., et al.: Integrating scale out and fault tolerance in stream processing using operator state management [C/OL]. In: Proceedings of the 2013 ACM SIGMOD International Conference on Management of Data, New York, NY, USA, pp. 725–736 (2013). https://doi.org/10.1145/2463676.2465282

17. Zhai, Y., Xu, W.: Efficient bottleneck detection in stream process system using fuzzy logic model [C]. In: 2017 25th Euromicro International Conference on Parallel, Distributed and Networkbased Processing (PDP), pp. 438–445 (2017)

18. Li, J., Pu, C., Chen, Y., et al.: Enabling elastic stream processing in shared clusters. In: 2016 IEEE 9th International Conference on Cloud Computing (CLOUD), pp. 108–115 (2016)

19. Heinze, T., Pappalardo, V., Jerzak, Z., et al.: Auto-scaling techniques for elastic data stream processing. In: 2014 IEEE 30th International Conference on Data Engineering Workshops, pp. 296–302 (2014)

20. Xu, L., Peng, B., Gupta, I.: Stela: enabling stream processing systems to scale-in and scale-out on-demand [C]. In: 2016 IEEE International Conference on Cloud Engineering (IC2E), pp. 22–31 (2016)

21. Frank, R., Castignani, G., Schmitz, R., et al.: A novel eco-driving application to reduce energy consumption of electric vehicles [C]. In: 2013 International Conference on Connected Vehicles and Expo (ICCVE), pp. 283–288 (2013)

22. Gulisano, V., Jimenez-Peris, R., Patino-Martinez, M., et al.: StreamCloud: an elastic and scalable data streaming system [J]. IEEE Trans. Parallel Distrib. Syst., pp. 2351–2365 (2013)

23. Apache ZooKeeper [EB/OL]. http://zookeeper.apache.org. Accessed 02 Apr 2020

24. Redis [EB/OL]. https://redis.io. Accessed 02 Apr 2020

25. Json-Data-Generator [EB/OL]. https://github.com/acesinc/json-data-generator. Accessed 02 Apr 2020

Chapter 3
Resource-Constrained Offloading in Edge Computing

Abstract As edge computing evolves, the need to process data at the edge of the network becomes more pressing, driven by the demand for real-time data processing and improved user interactions. However, this shift presents significant challenges, particularly in terms of resource constraints and privacy concerns. This chapter provides an overview of resource-constrained offloading, followed by offloading strategies, and potential scenarios. A computation model for local execution and task offloading is presented. Furthermore, the chapter reviews existing studies on task dependency, offloading to fog devices, dynamic offloading frameworks, and centralized versus distributed task offloading. Finally, the chapter highlights the challenges associated with resource-constrained offloading.

Keywords Resource-constraints · Task dependency · Offloading frameworks · Real-time processing · Task offloading · Fog devices · Privacy concerns

3.1 Overview of Resource-Constrained Offloading

Offloading refers to the process of transferring computational tasks or workloads from one device or system to another. In the context of edge computing, offloading typically involves moving computational tasks from resource-constrained devices (such as mobile devices or IoT devices) to more powerful and capable edge servers or cloud resources to reduce the latency in general and provide the application reliability in overall. Offloading can be in the form of computational offloading when the processing resources are limited, storage offloading when the edge devices have low storage capacity, network offloading when some application requires high bandwidth. However, the term is mostly tied with computational offloading.

To improve the performance of an IoT application, conserve the energy on the low end devices and improve the deadline meetup by reducing the latency offloading is a blessing for the low end resource constraint device. Offloading moves the computational tasks from the device generating the data to a more capable edge server or cloud resource. This can be partial (only certain tasks) or full (entire applications). Resource-constrained offloading in edge computing involves delegating computational tasks from devices with limited resources, such as IoT devices, wearables, and smartphones, to more capable edge servers or cloud resources. These resource-constrained devices often lack the necessary processing power, memory, and battery life to handle intensive computational tasks efficiently. By offloading tasks, these devices can operate more efficiently, conserve energy, and extend their operational lifespan. For instance, a smartphone might offload complex image processing or machine learning tasks to a nearby edge server, allowing the device to perform more efficiently without draining its battery.

This offloading approach not only improves the performance of individual devices but also enhances the overall system's responsiveness and reliability. It reduces latency by processing data closer to its source and optimizes bandwidth usage by minimizing the amount of data that needs to be transmitted to centralized cloud servers. However, it also requires careful management of network connectivity, security, and interoperability to ensure that offloaded tasks are handled seamlessly and securely. Resource-constrained offloading is crucial for applications requiring real-time data processing, such as augmented reality, smart health monitoring, and autonomous vehicles, where timely and efficient data processing can significantly impact user experience and system functionality.

3.1.1   Offloading Strategy

In the realm of computational resource management, offloading strategies have become integral to optimizing performance, efficiency, and resource utilization. The following methods are particularly significant in the context of mobile and edge computing, where devices such as smartphones, IoT gadgets, and edge servers may have limited processing power, memory, or battery life.

- **Local offloading:** Offloading can be performed to some powerful device or some other edge server available locally in the edge network.
- **Remote offloading:** Tasks can be offloaded to remote cloud computing servers, where ample resources are available. Usually this offloading is performed when sufficient resources are not available inside the edge network.
- **Partial offloading:** Only some tasks of an application or some services are offloaded to some other computing devices. This strategy is good for latency oriented applications as maximum computations will be performed inside the edge network hence meeting the deadlines.
- **Full offloading:** A complete application will be offloaded, this strategy is good for resources constrained edge devices, especially when resources like battery is draining, or in case of mobility when the edge device is moving away from the edge network. However, this will increase the latency.

3.1.2 Offloading Scenarios

The execution of computational tasks can occur at multiple locations, such as IoT devices, edge servers, or the cloud infrastructure. The selection of the offloading destination is dependent upon a delicate balance of various objectives and influencing factors. This section intends to provide an in-depth understanding of common offloading scenarios within the sphere of edge computing. Figure 3.1 depicts these scenarios by categorizing them into three layers based on their destination: the IoT devices, edge computing, and cloud computing layers, respectively.

- **IoT device-to-edge server:** The majority of IoT devices grapple with constraints in their computational capacities, rendering them inadequate for handling intricate tasks. Consider machine learning (ML), which demands substantial computing resources, particularly for training and inference involving Deep Neural Networks (DNN). Devices with restricted resources, such as smartphones, often face challenges when it comes to executing such tasks smoothly. Consequently, with cloud computing, computational tasks must be sent to the cloud server for processing. However, this approach incurs heightened costs associated with bandwidth and transmission latency, making it difficult to meet real-time demands, especially in industries with high requirements for swift processing. As depicted in Fig. 3.1, IoT devices have the ability to offload computational tasks to nearby edge servers, including Fog nodes, Cloudlets, or MEC servers, where

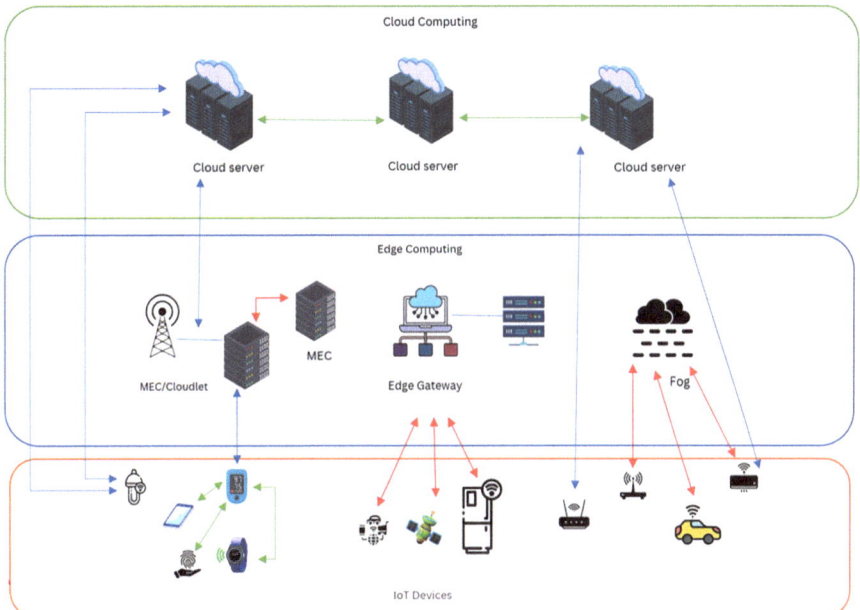

Fig. 3.1 Offloading scenarios

these tasks undergo further processing and analysis. This methodology effectively reduces costs and latency. Researchers have harnessed edge computing to offload DNN computational tasks from IoT devices to adjacent generic edge servers, such as MEC or cloudlets. This not only lightens the load on embedded devices but also results in significant improvements in application performance.

- **Offloading among IoT devices:** In today's modern era, IoT devices have seamlessly become an indispensable part of our daily lives, playing the roles of data generators and data consumers. Within the domain of edge computing, a substantial portion of IoT-generated data is managed at the network's edge, eliminating the need for transmission to distant cloud servers. Devices such as smartphones, tablets, and laptops are equipped with ample computing resources, many of which often remain untapped. When a mobile device encounters a demanding task that outdoes its own capabilities, it has the option to break down the application into smaller tasks and delegate them to nearby mobile devices with unused computing resources. This approach effectively lessens the resource limitations of individual devices and enhances overall resource utilization within edge environments.

- **Edge server-to-IoT device offloading:** In some specific scenarios, IoT devices necessitate external data, which may include temperature, humidity, and various environmental parameters, to perform their computations. Nevertheless, their limited functionality and resources bring challenges when it comes to measuring and storing big amounts of such data. In the context of cloud computing, IoT devices typically resort to requesting this information from the cloud. However, frequent cloud access can be both economically burdensome and tiring on the cloud server. Therefore, an efficient solution lies in leveraging distributed edge servers. These edge servers are strategically positioned in close proximity to IoT devices, boasting substantial computational capabilities and the capacity to collect data from multiple sources. Through the utilization of edge servers, the load on IoT devices, transmission delays, and the expenses associated with cloud connectivity can be greatly minimized.

- **Edge server-to-cloud offloading:** Edge servers like Cloudlets, Mobile Edge Computing systems, and Fog come equipped with substantial computational power and ample storage capacity, rendering them well-suited to efficiently manage the majority of tasks at the edge of the network. But, there are situations where they still depend on cloud services for data storage and accessibility. Take healthcare systems, for instance, which routinely transfer daily patient logs to the cloud for careful monitoring. Similarly, logs monitoring driver behavior often find their way to the cloud for storage and comprehensive analysis, enhancing the accuracy of vehicle insurance assessments. Additionally, in scenarios involving distributed resource deployments, edge servers, and cloud services collaborate to accomplish specific goals. Imagine, during snow disasters or forest fires, a fog system collecting real-time monitoring data from drones and transmitting it to the cloud for disaster assessment. In cases of severe scene damage, fog computing relies on the cloud's original geographic information to guide drones in conducting on-site search and rescue missions. As a result, the partnership between fog and cloud computing is marked by task allocation and collaborative efforts, ultimately leading to the provision of comprehensive services.

- **IoT device-to-cloud offloading:** In certain scenarios, IoT devices may encounter situations where they require the assistance of remote cloud servers to manage tasks related to data storage or computation. Mobile Cloud Computing (MCC), as explored in research, enables mobile devices to offload resource-intensive tasks to powerful centralized cloud platforms such as Amazon EC2, MS Azure, and Google Cloud. These cloud services also offer extensive data storage capabilities. MCC brings forth numerous benefits, including the extension of battery life, the empowerment of advanced applications, and the provisioning of extensive data storage options for mobile users. Another framework, CloneCloud, simplifies the process of cloning specific segments of an application, which can then be offloaded to the cloud. When the application reaches the cloned segment, the application thread seamlessly transitions from the mobile device to its cloud clone. It can subsequently be re-offloaded to the mobile device at the program's current execution point to resume processing. CloneCloud effectively reduces the overhead allied with application execution.

- **Offloading among edge servers:** Edge computing provides substantial benefits to IoT devices, including heightened computational power and swift response rates. Since the computational capabilities of individual edge servers have their limits. Hence, the necessity arises to incorporate numerous edge nodes to uphold load equilibrium and enable seamless data exchange, especially in support of collaborative services like cooperative edge computing.

  To illustrate this, consider the context of coronavirus prevention. Through collaborative edge, hospitals can share crucial information like the number of infected individuals, symptoms exhibited, and available beds. Simultaneously, the disease control centers can utilize the collaborative edge information to probe and monitor coronavirus contacts, promptly updating the contact information. Communities can leverage collaborative edge to take necessary actions, such as isolating individuals who have come into contact with the virus. Collaborative edge enables disease control centers to coordinate with hospitals, communities, and pharmaceutical companies, effectively scheduling resources and enhancing efficiency in prevention and control efforts. By utilizing a collaborative edge, various stakeholders can work together to optimize resource utilization and elevate the overall effectiveness of coronavirus prevention measures.

- **Cloud-to-IoT device offloading:** It's relatively unusual for IoT devices to directly receive tasks from the cloud, there are specific scenarios where the cloud may necessitate the device or sensor to execute particular tasks and provide feedback on the results. In real-time video analysis, the response time plays a pivotal role in determining whether quality of service requirements are met. Due to prolonged data transmission latency and privacy concerns, traditional cloud computing is no longer suitable for video analysis. Instead, the cloud assigns the task of locating missing children to all devices within the specified area. Each surveillance camera, upon receiving the task, conducts a search within its local data and expeditiously relays the results back to the cloud.

This approach, in contrast to conventional cloud computing, harnesses edge computing, and parallel processing, resulting in markedly accelerated response times. Additionally, certain tasks, such as image, sound, and signal acquisition, must be offloaded from the cloud to the edge devices, as they rely heavily on the capabilities of the underlying equipment.

3.2   Computation Model for Local Execution and Task Offloading

The computation model can normally be constructed with the aim of determining various parameters, including execution and transmission times, energy usage, and cost, during the execution of a task. The task execution can take place on an SMD, referred to as local execution. Additionally, a task has the option to be offloaded for execution on an MEC server. The task execution on a MEC server comprises two distinct operations: task offloading and edge computing. The following subsection provides an overview of the general models used to calculate the latency during the execution of a task and the power usage of an SMD during both executions either locally on the MEC server.

Let's consider a set of N SMDs, represented as $\mathcal{A} = \{A_1, A_2, A_3, \ldots, A_N\}$, that are wirelessly connected to the resources located remotely. Each SMD has X tasks represented as $\mathcal{T} = \{\tau_{n,1}, \tau_{n,2}, \tau_{n,3}, \ldots, \tau_{n,X}\}$. For each SMD A_n two parameters (f_n, e_n) are used in the modeling process, where f_n is the computation ability of SMD, and e_n denotes the power used for task transmission when offloaded. Further, for each task $\tau_{n,x}$ two parameters $(d_{n,x}, c_{n,x})$ are used in the modeling process, where $d_{n,x}$ is the size of input data $\tau_{n,x}$, and $c_{n,x}$ is the count of necessary CPU cycles to complete $\tau_{n,x}$.

If SMD possesses appropriate computational resources, which also include capacity of storage and CPU frequency, it will execute the task locally. In such cases, the SMD utilizes its own resources to complete the task without involving external entities. However, if the SMD lacks the necessary resources, it turns to offload the task to a remote resource, typically an MEC server. The overall task completion time can be divided into three components: the time spent on local processing, offloading, or task transfer time, and the time required for task execution on MEC server. During either task processing locally or task transfer to MEC server, the SMD consumes energy. The subsequent explanation outlines the process of calculating latency and energy consumption in both local and edge computing contexts.

- **Local computing.**   In on-premises computing, the task execution time $\tau_{n,x}$ depends on the number of CPU cycles $c_{n,x}$ required, and the CPU frequency f_n of an SMD and can be computed as:

$$t_{n,x}^{local} = \frac{c_{n,x}}{f_n} \tag{3.1}$$

The energy consumed by an SMD during the execution of task $\tau_{x,n}$ is expressed as:

$$e_{n,x}^{local} = \kappa c_{n,x} f_n^2 \qquad (3.2)$$

here κ represents the switched capacitance factor and is contingent on the chip's design. Different studies have recommended varying values for the coefficient κ.

- **Task offloading.** Offloading a task wirelessly through an Access Point involves various factors that determine the transmission rate. These factors include the wireless channel's bandwidth, the SMD's transmission power, the distance between the SMD and the Access Point, and the interference present among different channels. Each of these elements plays a key role in shaping the efficiency and speed of the task-offloading process. The transmission rate can be defined by Shannon-Hartley's theorem:

$$R_n = B \log_2 \left(1 + \frac{e_n G_n}{\sigma + \sum_{i \neq n, j \neq n} e_{i,j} G_{i,j}} \right) \qquad (3.3)$$

where B is the bandwidth of the channel, e_n signifies the transmission energy employed by A_n to send a task to an edge server via the access point, G_n denotes for the wireless channel's gain connecting A_n and the relevant access point.
$G_n = \left(ed_{n,a}^{-\eta} \right) \mid h_{n,a} \mid^2$, where $ed_{n,a}^{-\eta}$ represents the attenuation due to signal propagation, $ed_{n,a}$ shows the Euclidean distance between A_n and access point, $\eta \geq 2$ is the path loss exponent, $h_{n,a}$ is the respective Rayleigh fading channel coefficient that adheres the distribution of $N(0, 1)$, and σ indicates the background interference.

Usually, the time it takes to transmit a task from an SMD to MEC server is predominantly influenced by wireless communication, given the significantly broad bandwidth of the wired channel, resulting in a task transmission time that is almost negligible. Hence, the transmission time of $\tau_{n,x}$ can be computed as:

$$t_{n,x}^{trans} = \frac{d_{n,x}}{R_n} \qquad (3.4)$$

where $t_{n,x}^{trans}$ represents the duration dedicated to transmitting the input data for $\tau_{n,x}$ from A_n to the MEC server.

The energy utilization of A_n for task offloading $\tau_{m,n}$ to the MEC server is specified as:

$$e_{n,x}^{trans} = e_n t_{n,x}^{trans} \qquad (3.5)$$

where e_n represents the transmission power of A_n.

- **MEC Server.** Upon task arrival, the MEC server allocates computation capability $f_{n,x}^{edge}$ to the associated SMD for task execution. Consequently, the task completion time at the MEC server can be described as follows:

$$t_{n,x}^{MEC} = \frac{c_{n,x}}{f_{n,x}^{edge}} \tag{3.6}$$

where $\sum_{n=1}^{N} \sum_{x=1}^{X} Z_{n,x} f_{n,x}^{edge} \leq F$, which implies that the accumulation of assigned computation capabilities should be in line with the MEC server's total computational capacity.

The average time T and energy consumption E of an SMD to complete a task can be expressed as:

$$T = \frac{1}{NX} \sum_{n=1}^{N} \sum_{x=1}^{X} (1 - Z_{n,x}) t_{n,x}^{local} + Z_{n,x} \left(t_{n,x}^{trans} + t_{n,x}^{MEC} \right) \tag{3.7}$$

$$E = \frac{1}{NX} \sum_{n=1}^{N} \sum_{x=1}^{X} (1 - Z_{n,x}) e_{n,x}^{local} + Z_{n,x} e_{n,x}^{trans} \tag{3.8}$$

where $Z_{n,x}\{0, 1\}$ is a binary value, characterized as $Z_{n,x} = 0$ when the task is executed locally; otherwise, $Z_{n,x} = 1$.

This brings us to a multiobjective optimization challenge where the aim is to minimize the following:

$$P1 : \min_{Z_{n,x}} \{E, T\}$$

s.t

$$C1 : Z_{n,x} \in \{0, 1\}$$

$$C2 : \sum_{n=1}^{N} \sum_{x=1}^{X} Z_{n,x} f_{n,x}^{edge} \leq F$$

While defining this problem, it is conceivable to introduce additional constraints, such as the necessity for latency to be shorter than each task's deadline. To simplify this multiobjective optimization problem, it can be converted into a single-objective optimization task by allocating weights to each objective function: $O = \gamma T + (1 - \gamma)E$

$$P2 : \min_{Z_{n,x}} \{O\}$$

s.t.

$$C1 : Z_{n,x} \in \{0, 1\}$$

$$C2 : \gamma \in \{0, 1\}$$

$$C3 : \sum_{n=1}^{N} \sum_{x=1}^{X} Z_{n,x} f_{n,x}^{edge} \leq F$$

where $\gamma \in [0, 1]$ is the weight value. The determination of γ varies with the type of application in consideration. For instance, γ is typically adjusted to be near 1 for time-critical applications and near 0 for energy-centric applications.

3.3   A Review of Existing Studies on Task Offloading

- **Task's Dependency:** Prior studies have assumed that an application's tasks are independent to simplify the offloading process. However, tasks in many applications are dependent. One task may have a predecessor task that ensures the beginning of the task or there may be a successor task that will be continued after the completion of the previous task. Assuming the interdependence of the tasks, some authors used the call gap between the methods of application which indicates dependent tasks [1]. The interdependence of the tasks will make scheduling more difficult [2, 3].

    All of the tasks are completed in a sequence, and the output data produced by one task serves as the input for the following one in [4] whereas the authors arranged all tasks in a priority queue based on their assignment urgency instead of executing in sequence in [5]. They focus on designing a cost-efficient system for offloading tasks for multi-users. To minimize the system cost under constraints, they used the heuristic offloading algorithm. Following the first assignment, they transferred the tasks based on the relative remaining energy from high-cost devices to low-cost devices in an effort to reduce system costs while meeting the energy and completion time limitations.

    A method for offloading independent tasks in a cloud-edge environment to reduce the total amount of time required for latency-sensitive applications is suggested in [6]. They employ the fuzzy logic technique to schedule tasks. They presented an algorithm for task scheduling. It compares the capacity of the resource with predefined values and offloads the tasks based on those values. Simulation for this approach shows that the overall latency time is decreased but the paper only focuses on offloading the independent tasks. Moreover, no algorithm or architecture is proposed for offloading the tasks of AR/VR or video games.

- **Offloading to Fog Devices:** The authors have discussed the limited computation capacity of fog nodes in fog computing where the single IoT device may try to offload its tasks to multiple fog nodes in [7]. The authors have applied the down-link Non-Orthogonal Multiple Access NOMA) to make the offloading efficient. By this technique, the single IoT device can offload its tasks simultaneously to multiple fog devices. But they only offload the tasks to the fog devices while offloading to the CPU, cloud server, and edge devices are not considered, while in [8], uplink NOMA is used for computation offloading, where several mobile users simultaneously offload tasks to a fog node. However, it cannot be applied for offloading tasks from an IoT device to multiple fog nodes. In addition, there are two issues, the computation capacity of a fog node is unlimited (as assumed in work), which may not be practical. Moreover, it cares only about the profit of executing a single task, i.e., the short-term profit.

- **Dynamic Offloading:** The developers of MAUI [9] presented a dynamic offloading framework with method-level (i.e., application component-level) granularity. A mobile cloud computing framework with dynamic offloading was also introduced by ThinkAir [10]. The suggested framework creates a virtual machine copy of the smartphone platform (VM). It provides a library and a compiler to make the adaptation of games simple. The wrappers and utility functions are generated by a code generator. The ARM-based instructions of the remote methods are converted into x86 instructions using a customized native development kit (NDK). The four objective functions that ThinkAir defines combine execution time, energy use, and financial cost.

  F. Messaoudi et al. [11], presented an offloading method to enhance the functionality of "Unity 3D". They offload the GO (Game Objects) to the server. The frames of the GO are rendered on the server using a heuristic algorithm and the game is played. Then the changing frames of the GO are rendered back to the mobile device. But this paper only focuses on increasing the game capacity to run complex games while rendering activity and increased latency are not discussed. For decreasing latency, the authors in [12] reviewed the most important algorithms used in the resource allocation management process at the MEC, which are the DPSO, ACO, and basic PSO. The experiments prove that the DPSO, a task migration algorithm, is the better and more appropriate algorithm for applications requiring low latency. It keeps track of the device positions (mobility) and then tasks are offloaded to the immobile devices.

- **Centralized and Distributed Task Offloading:** Centralized task offloading and distributed task offloading are the two categories of task offloading in Mobile Edge Computing (MEC). For the former [13, 14], in order to maximize the task completion ratio, Baron et al. [13] suggested a multi-user task offloading strategy among many edges. To reduce overall energy consumption, Jiang et al. [14] advised a centralized job offloading solution based on deep learning and MEC. The satisfaction of a user with the offloading decisions is never taken into account by centralized techniques. For the latter [15, 16]. For industrial IoT-Edge-Cloud computing scenarios, Hong et al. [15] propose multi-hop cooperative computation offloading. Deep reinforcement learning is used by Wang et al. [16] to study the offloading problem and resource allocation. Although it is a more realistic scenario, most existing research neglect to take into account the effects of offloading among edge nodes.

  A three-way round-robin game involving users, edge nodes, and service providers was proposed by Ma et al. [17] as part of their design of a service-oriented resource allocation method. In order to solve the joint wireless and computational resource allocation challenge, Gu et al. [18] suggested a matching game-based student project allocation game strategy. These studies, however, only take into account the issue of how users and edge servers should be allocated resources; they do not address the issue of offloading strategy between edge devices.

3.4 Challenges for Resource-Constrained Offloading

Computation offloading is key in MEC networks for reducing latency, enhancing battery life, and improving user QoE. Although significant progress has been made, numerous challenges still exist that require further investigation. This section highlights the major challenges, with a particular emphasis on addressing security and privacy concerns.

- **Handling task inter-dependencies.** Due to the computational constraints, larger tasks are often broken down into interconnected subtasks that are offloaded to various edge servers. Multiple edge servers then work together to benefit users. The subtasks are dependent on one another, meaning the input of previous task relies on the output of the next, which makes resource scheduling more challenging when trying to minimize the total task completion time. Resource allocation for these subtasks must consider both the amount of allocated resources and the timing of these allocations. Considering the interdependence of subtasks within a task, the inefficiency from allocating resources for the entire task duration can not be ignored. It is crucial to optimize resource utilization and minimize task completion time to benefit both users and infrastructure providers.

- **Optimal server selection.** Due to growing diversity of service offerings in MEC networks, it is impractical and unnecessary to equip each edge server with infinite resources and hardware capable of handling every type of task, which in turn minimizes the scalability of MEC networks. Network Functions Virtualization (NFV) offers an effective solution by using virtualization technology to launch services and applications. Each edge server's services are constrained by the functions it supports, which can lead to some edge servers being unavailable for certain tasks. Therefore, it is essential to consider the types of edge servers as well as the availability of resources when offloading computation tasks for users. Conventional NFV approaches only account for tasks with chain structures, overlooking the potential graph features of tasks in new applications, such as VR/AR, and other intelligent applications. When tasks have QoS constraints and require multiple resources, selection of edge servers for different subtasks based on server types becomes more challenging and is a crucial area for further investigation.

- **Current environmental state observation.** As user tasks are offloaded to edge servers, the QoS of users is substantially impacted by the dynamic nature of

MEC networks. The transmission rate of wireless links is particularly affected by the bandwidth, the transmitting power of base stations (BSs), and variable conditions of wireless channel. Conventional approaches, including those based on game theory and heuristic algorithms, lack the flexibility to adjust computation offloading in response to the fluctuating conditions of MEC networks. These methods often set channel attenuation and other parameters to constant values in simulations. However, the variability of the wireless channel can be mitigated by calibrating the transmission power of BSs. Consequently, to address random wireless channels within power budget limits, optimizing both BS power control and user computation offloading is crucial for practical implementation in MEC networks.

- **Privacy concerns.** Privacy preservation during task offloading in the edge computing environment, is a significant but less explored issue. There is a risk of privacy leakage in numerous ways, especially when computing tasks that involve user's private information are delegated to the edge server. For instance, an attacker may obtain the user's private information by observing the user's offloading behavior stats because a user's offloading pattern is strongly tied to their channel state and usage patterns. Furthermore, because offloaded tasks are transferred wirelessly, information may be intercepted by eavesdroppers. A compromised or untrusted edge server can also result in a violation of user privacy. Users may suffer severe safety problems as well as revenue losses as a result of this. Therefore, it is crucial to ensure privacy during the offloading process. The subsequent Chap. 4 covers privacy-preserving offloading, followed by our approach for addressing privacy and latency concerns in edge cloud environments, and presents our findings.

References

1. Zhang, Y., Liu, H., Jiao, L., Fu, X.: To offload or not to offload: An efficient code partition algorithm for mobile cloud computing. In: Proc. IEEE 1st Int. Conf. Cloud Netw. (CLOUDNET), Nov. 2012, pp. 80–86 (2012)
2. Zhang, W., Wen, Y., Wu, D.O.: Energy-efficient scheduling policy for collaborative execution in mobile cloud computing. In: Proc. IEEE Conf. Conf. Comput. Commun. (INFOCOM), Apr. 2013, pp. 190–194 (2013)
3. Jia, M., Cao, J., Yang, L.: Heuristic offloading of concurrent tasks for computation-intensive applications in mobile cloud computing. In: Proc. IEEE Conf. Comput. Commun. Workshops (INFOCOM WKSHPS), Apr./May 2014, pp. 352–357 (2014)
4. Abdel-Jabbar, M.A.H., Kacem, I., Martin, S.: Unrelated parallel machines with precedence constraints: Application to cloud computing. In: Proc. IEEE 3rd Int. Conf. Cloud Netw. (CloudNet), Oct. 2014, pp. 438–442 (2014)
5. Fan, Y., Zhai, L., Wang, H.: Cost-efficient dependent task offloading for Multiusers. IEEE Access **7**, 115843–115856 (2019)
6. Almutairi, J., Aldossary, M.: A novel approach for IoT tasks offloading in edge-cloud environments. Journal of Cloud Computing 10(1), 28 (2021)
7. Wei, Z., Jiang, H.: Optimal offloading in fog computing systems with non-orthogonal multiple access. IEEE Access **6**, 49767–49778 (2018)

8. Wang, F., Xu, J., Ding, Z.: Optimized multiuser computation offloading with multi-antenna NOMA. In: Proc. IEEE Global Commun. Conf. (GLOBECOM) Workshop, Singapore, Dec. 2017, pp. 1–7 (2017)

9. Cuervo, E., Balasubramanian, A., Cho, D., Wolman, A., Saroiu, S., Chandra, R., Bahl, P.: MAUI: making smartphones last longer with code offload. In: Proceedings of the 8th International Conference on Mobile Systems, Applications, and Services (MobiSys 2010), San Francisco, California, USA, June 15–18, 2010, pp. 49–62 (2010)

10. Kosta, S., Aucinas, A., Hui, P., Mortier, R., Zhang, X.: Thinkair: Dynamic resource allocation and parallel execution in the cloud for mobile code offloading. In: Proceedings of the IEEE INFOCOM 2012, Orlando, FL, USA, March 25–30, 2012, pp. 945–953 (2012)

11. Messaoudi, F., Ksentini, A., Bertin, P.: Toward a mobile gaming based-computation offloading. In: 2018 IEEE International Conference on Communications (ICC), pp. 1–7 (2018). https://doi.org/10.1109/ICC.2018.8422518

12. Alrazgan, M.: Internet of medical things and edge computing for improving healthcare in Smart Cities. Math. Prob. Eng. Hindawi **2022**, 5776954 (2022)

13. Baron, B., Spathis, P., Rivano, H., de Amorim, M.D., Viniotis, Y., Ammar, M.H.: Centrally controlled mass data offloading using vehicular traffic. IEEE Trans. Netw. Serv. Manag. **14**(2), 401–415 (2017)

14. Jiang, F., Ma, R., Sun, C., Gu, Z.: Dueling deep q-network learning based computing offloading scheme for f-ran. In: 2020 IEEE 31st Annual International Symposium on Personal, Indoor and Mobile Radio Communications, pp. 1–6 (2020)

15. Hong, Z., Chen, W., Huang, H., Guo, S., Zheng, Z.: Multi-hop cooperative computation offloading for industrial IoT–edge–cloud computing environments. IEEE Trans. Parallel Distrib. Syst. **30**(12), 2759–2774 (2019)

16. Wang, Y., Ge, H., Feng, A., Li, W., Liu, L., Jiang, H.: Computation offloading strategy based on deep reinforcement learning in cloud-assisted mobile edge computing. In: 2020 IEEE 5th International Conference on Cloud Computing and Big Data Analytics (ICCCBDA), pp. 108–113 (2020)

17. Ma, S., Guo, S., Wang, K., Jia, W., Guo, M.: A cyclic game for service-oriented resource allocation in edge computing. IEEE Trans. Serv. Comput. **13**, 723–734 (2020)

18. Gu, Y., Chang, Z., Pan, M., Song, L., Han, Z.: Joint radio and computational resource allocation in IoT fog computing. IEEE Trans. Veh. Technol. **67**, 7475–7484 (2018)

Chapter 4
Privacy Preserving Offloading

Abstract While much research has been conducted on privacy preservation in conventional cloud computing, these techniques may not be straightforwardly applied to offloading edge computing. The reason is that conventional cloud servers are typically employed by industry leaders like Alibaba and Amazon at a centralized level, having thorough and sophisticated security protocols with high integrity. Even, edge servers might be installed by various organizations within an open ecosystem, featuring comparatively less stringent security measures. This can render them less reliable than conventional cloud servers and more inclined to both cyber threats and physical breaches.

The privacy preservation in edge computing during data offloading has gained significant attention recently. A variety of research initiatives have been undertaken to probe various privacy preservation techniques in offloading, including data transfer metrics, wireless transmission techniques, and offloading destination selection schemes. In this chapter, we address privacy and latency issues in the context of offloading computations to the edge-cloud computing environment. Our solution focuses on leveraging the power of inductive learning to train a feature extractor and a centralized neural network, all while preserving the integrity of sensitive data at the network's edge. We leverage local differential policy approach to ensure that private data is retained locally and never sent to the cloud. Additionally, our approach factors in data transmission costs and the resources available on edge devices, with the extensive aim of optimizing privacy and efficiency within mobile edge intelligent systems.

Keywords Inductive learning · Security · Local differential privacy · Edge-cloud integration · Privacy preservation · Cyber threats · Security measures · Data transmission

4.1 Privacy-Preserving Offloading in Edge Intelligence Systems with Inductive Learning and Local Differential Privacy

4.1.1 Introduction and Motivations

Phones, smartwatches, and other modern electronic devices generate a large volume of data at the network's edge, which is essential for training AI models. However, training a DNN on mobile or IoT devices is challenging due to their limited processing capabilities. Traditionally, edge devices offload their data to cloud servers for processing, but this approach faces challenges like latency, storage costs, and privacy concerns due to the increasing data volume. These concerns include the risk of data exposure to internal and external threats and potential information leakage even from anonymized data [1–3]. To address these issues, a new paradigm has emerged, shifting some computing tasks from the cloud to edge devices, known as edge computing and edge intelligence when AI is leveraged in edge computing. This approach employ the division of DNNs into multiple components to solve the mentioned challenges.

The widely adopted approach is similar to transfer learning [4, 5], using a pre-trained model or training with auxiliary or public datasets. The pre-trained model is divided into two parts: low-level layers that reduce input dimensionality and are transferred to edge devices as feature extractors. These extractors compress data and reduce network latency. A differentially private algorithm can protect sensitive information during feature extraction. The server side uses these features to fine-tune the model's second half. State-of-the-art frameworks [6–9] use this method, relying on auxiliary datasets for training. However, many frameworks focus on image data, and finding publicly available datasets for other fields, such as IoT, is challenging due to privacy concerns. IoT devices track personal data, making public datasets rare. This chapter introduces a new learning system that preserves privacy and reduces latency by using only the sensitive data from edge devices, eliminating the need for auxiliary datasets.

Deploying the full AI capacity on edge devices is challenging due to their limited resources, and service providers may be reluctant to deploy fine-tuned models on edge devices for copyright reasons. Additionally, a centralized server might be needed to generate the final result for end-users, making it impractical to train or deploy third-party cloud service provider models on edge devices, as suggested by Federated Learning [10]. Furthermore, clients may not want to release their sensitive data to the cloud service provider. To address these conflicting needs, we propose a new hybrid learning approach through a framework called DeepGuess, illustrated in Fig. 4.1. Our framework utilizes the AutoEncoder (AE) architecture to reduce network latency and provide an initial layer of privacy. On IoT devices, the Encoder acts as a feature extractor, identifying key features from sensitive data. To add a second privacy layer, we introduce a differential privacy algorithm that injects random noise into the features on the edge side before sending them to the

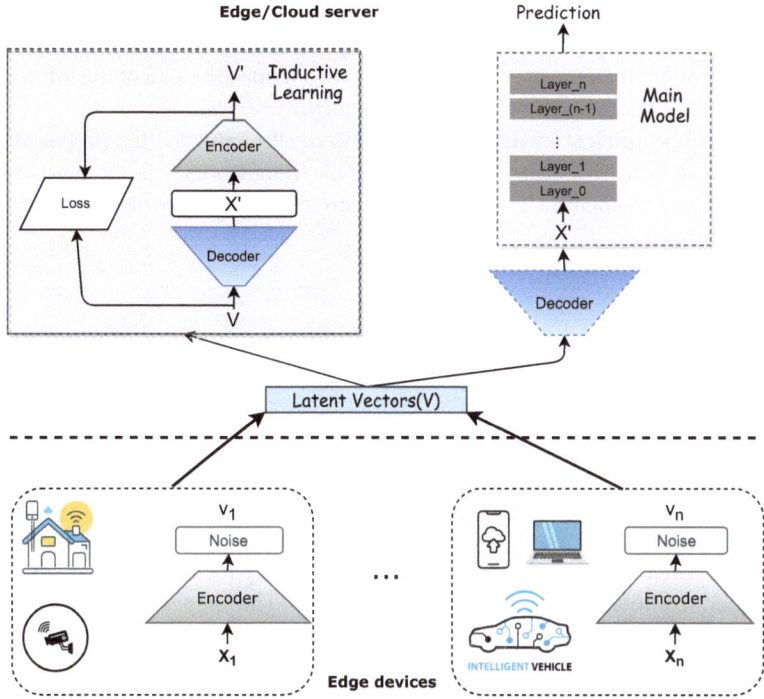

Fig. 4.1 The Encoder, Decoder, and Main model are all initialized randomly. Important features are obtained from edge side sensitive data, and noise is added before being transferred to the server

central server for further processing. A major challenge is that the Encoder used as the client-side feature extractor is not pre-trained. The extracted features, referred to as a latent vector, will be discussed further in the chapter. The contributions of our work are listed as follows:

1. **We designed inductive learning:** A new learning process that does not rely on auxiliary datasets to train the client side feature extraction module. Our system's unique data source is edge devices. To access edge data in a privacy-preserving manner, introduce inductive learning. The proposed solution ensures that the raw private data is not centralized on the central server while providing the data utility.
2. **A new standardized approach for enabling ϵ-Local Differential Privacy:** Transferring extracted features from a sensitive dataset to the server does not guarantee total privacy since it has been shown possible to leak sensitive information. To provide a rigorous privacy guarantee, we propose a new mechanism to perturb the extracted features using differential privacy.
3. **Reduced computation on edge devices and low latency:** The edge device only conducts forward passes using the Encoder. The entire training takes place

in the cloud. As most heavy computations migrate to the cloud, we alleviate the challenges of limited resources on edge devices. By transferring the latent features rather than the user's raw data, we also reduce the size of the information transferred over the network hence the latency.

4. **Thorough empirical evaluation:** We empirically validated the proposed solution via rigorous experiments against SOTA frameworks. The obtained results demonstrate DeepGuess's robustness and effectiveness in lowering communication overhead.

4.1.2   Data Anonymization Mechanisms

Anonymization is a procedure that disguises or modifies information in datasets, making it anonymous. As a result, instances of the derived dataset can no longer be associated with a specific identity.

Basic Strategies The basic anonymization strategies attempt to eliminate all sensitive attributes before the dataset is released. They include data masking [11], data generalization [12], and data swapping [13]. The biggest drawback of such simple anonymization methods is the possibility of identifying records from the anonymized data using cross-referencing techniques [14–16]. The second concern is that these strategies are often conducted on the server side, so users must rely on service providers' goodwill to anonymize their data.

Noise Addition Noise addition techniques work by using a stochastic number to alter the value of sensitive attributes in the dataset [17, 18]. The state-of-the-art anonymization strategy is the Differential Privacy (DP) introduced by Dwork [2]. DP is the only strategy to have formal verification of data confidentiality with mathematical proofs [19, 20]. Formal verification is essential and makes it possible to quantify the re-identification risk of data records. The DP algorithm incorporates random noise into the data so that anything tampered by the adversaries becomes perturbed, hence imprecise, making it far more difficult to violate privacy.

Differential Privacy (DP) and Local Differential Privacy (LDP) In DP, the amount of noise applied to the data is primarily controlled by ε. $\varepsilon = 0$ is the maximum noise and guarantees perfect privacy. $\varepsilon = +\infty$ is the lowest noise and does not guarantee any privacy. ε is called the privacy budget. The concept of ε−Differential Privacy ($\varepsilon - DP$) was introduced in [1], and it is formalized as follows:

Definition 4.1 ($\varepsilon - DP$) Let $\varepsilon >= 0$ and A be a randomized algorithm that takes a dataset as input, and A' the image of A. The algorithm A is supposed to provide

$\varepsilon - differential\ privacy$ if, for any adjacent datasets D_1 and D_2 that differ on a single element, and any subsets S of A':

$$Pr[A(D_1) \in S] \leq \exp(\varepsilon).Pr[A(D_2) \in S]$$

where $Pr[A(D_1) \in S]$ indicates the probability that the outputs of algorithm A belong to S.

The problem with the $\varepsilon - DP$ is that it remains centralized. Therefore, a new DP variant called Local Differential Privacy (LDP) was proposed. LDP allows each client to add noise to the sensitive information locally. The concept was introduced in [21], and it is formalized as follows:

Definition 4.2 ($\varepsilon - LDP$) Let $\varepsilon >= 0$ and A be a randomized algorithm that takes its input in X with X representing the user's local data. A' is the image of A. The algorithm A is said to provide $\varepsilon - local\ differential\ privacy$ if and only if, $\forall x_1, x_2 \in X$ and $\forall y \in A'$:

$$Pr[A(x_1) = y] \leq \exp(\varepsilon).Pr[A(x_2) = y]$$

To know how much noise or randomness we can introduce with ε, it is important to estimate the data sensitivity. In DP, the global data sensitivity given by Eq. (4.1) is the maximum effect between two adjacent items or datasets (d_1, d_2) on the output of an arbitrary f function, typically referred to as the query.

$$S_f = max|f(d_1) - f(d_2)|_1 \tag{4.1}$$

Randomized Response: Coin Flips LDP is a recent technique, but the intuition behind it is quite old [22]. It was introduced to collect statistical data from users' answers while ensuring confidentiality. In a survey where a person has to answer YES or NO, the procedure is as follows: The user flips a coin in private if the head comes up, flips the coin again a second time, ignores the results, and answers truthfully. If the first flip was not a head, he flips the coin a second time and answers "Yes" if it is a head or "No" otherwise. The second flip of the first case is used to fool a watching stranger. Let us suppose p is the probability that the person will answer truthfully and $(1 - p)$ otherwise. This approach provides ϵ-DP for $p = e^\epsilon/(1+e^\epsilon)$ [23]. Google's LDP framework RAPPOR [24] also uses this process to collect Chrome users' data (home pages, Chrome configuration strings).

4.2 Proposed Solution

This section first provides an overview of our proposed framework (Sect. 4.2.1), including an illustration of how the inductive learning mechanism works

(Sect. 4.2.2). We also discuss the use of randomized unit response and the Laplace-DP mechanism to implement our framework's privacy-preserving mechanism (Sect. 4.2.4).

4.2.1 Framework Overview

Figure 4.1 provides the general structure for the DeepGuess framework. The framework comprises two parts, running respectively on the edge device and the server. The edge device runs the Encoder and the noise module, while the cloud server has the full AE (composed of the Encoder and Decoder) and the Main Model that we need to train for our task. During the training process, the AE and the Main model's weights are randomly initialized and fine-tuned using the data produced by edge devices.

AEs are particular types of neural networks that learn to output their input. They are commonly used to learn a latent representation of a dataset or as a dimensionality reduction technique. As shown in Fig. 4.2a, the AE takes advantage of DNN splitting [25] property and splits the network architecture into two parts: Encoder and Decoder. The Encoder is used to convert the input into a reduced latent representation. Alongside, the Decoder tries to restore from the reduced encoding a representation as close as possible to its original input. There is also another AE variant called Sparse AE [26] (SAE), which, instead of reducing the input dimensionality, will rather increase it (Fig. 4.2b).

The AE acts as a device-to-server data bridge. As its name implies, the Main model is used for the primary task. It infers a value of interest that can be used by a more complex subsystem to generate a final result for the client. Deployed at the network edge, the Encoder makes it possible to reduce the data to be transmitted to a noisy latent representation. Once this noisy latent representation has been transferred to the server, Fig. 4.3 further displays two possible configurations for training the whole system:

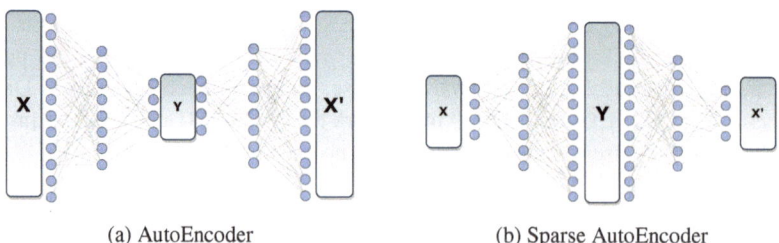

(a) AutoEncoder   (b) Sparse AutoEncoder

Fig. 4.2 AutoEncoder and Sparse AutoEncoder: Unlike traditional AutoEncoder, a Sparse AutoEncoder's latent vector Y size is bigger than the input X. (**a**) AutoEncoder. (**b**) Sparse AutoEncoder

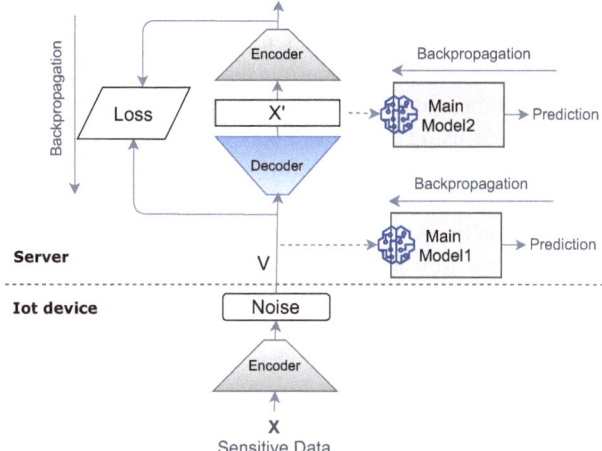

Fig. 4.3 Different framework configurations: The Main model's input may be the noisy latent vector or the reconstructed data

- The first is to use the latent vectors V as input to train the Main model. This configuration is the best given that the Encoder has already performed the feature extraction. The Main model could only focus on predicting the final output. This configuration then reduces the complexity of the Main model and leads to faster convergence.
- The Decoder in the second configuration takes the noisy latent V vectors and approximates the original data X'. X' is then used for training the Main model. This configuration will necessitate designing a more complex Main model, which may carry both feature extraction on X' and prediction of the final output, hence slower to train and converge than in the first case.

Consequently, the Encoder acts as a data compressor and the Decoder as a data decompressor, helping to considerably reduce the volume of data during network transmission and communication and latency costs. The inferences and the whole training process to tune the AE and main model weights occur in the cloud. The client's device performs minimal processing: i.e., only forward pass with Encoder and noise addition. We can also observe that the server never directly interacts with sensitive user data. As the Encoder reduces sensitive input to a lower dimension, information loss pushes the AE to prioritize and learn general aspects of the input during training rather than user-specific information. This results in an imperfect reconstruction on the server during the process of inferences. Additionally, adding noise to the client's latent vectors will make restoring sensitive information on the server challenging.

The scenario described in the introduction section, which consists of splitting a neural network into two parts and using the first as a feature extractor, has some limitations in edge computing. This approach works with one model architecture at a time and requires a public dataset to pre-train the feature extractor. A neural network design consists of many fine-tuning and often involves testing several architectures before selecting the correct one. Separating the feature extraction module as in our solution provides more flexibility as several networks of various tasks can be evaluated simultaneously using the same Encoder and Decoder. For the system to function properly, we must train the AE before deploying the encoder to the devices. Related research efforts have proposed two solutions. The first consists of using auxiliary datasets to train the AE. In the second option, we collect the edge data and centralize it on the server to train the AE. As mentioned earlier, we will not rely on auxiliary datasets for our framework, and the second option is not desirable due to privacy concerns. Deprived of both options, we have introduced a new learning mechanism called inductive learning, which is discussed in the following section. Using inductive learning allows the server to tune the AE with the latent vectors it receives from the edge devices.

4.2.2 Inductive Learning

Wireless or inductive charging allows charging a device's battery without plugging it into a power source. Similarly, inductive learning is a process that allows a neural network to learn the structure of a dataset without explicitly involving it in each backpropagation step. Learning a dataset structure is the principal purpose of AE. In the direct training mode illustrated in Fig. 4.4, the dataset (X) representing the input and labels is involved in the backpropagation process.

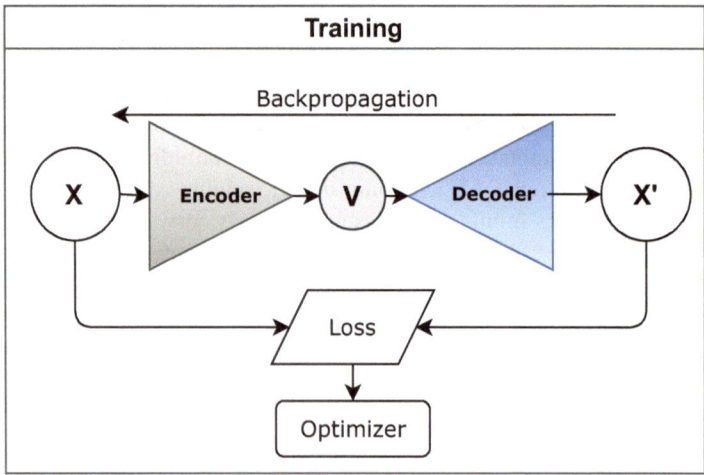

Fig. 4.4 Direct mode configuration

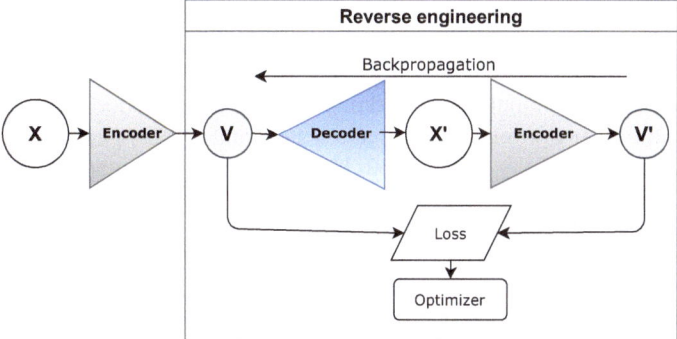

Fig. 4.5 Inductive learning configuration

The indirect training configuration adopted by inductive learning is displayed in Fig. 4.5. In this configuration, the dataset is not included in the backpropagation. It could be seen as a reverse engineering procedure in which, by just extracting certain key features V from X, we try to learn the structure of X.

During training, the AE is sparse, with the Decoder as the first component, the Encoder as the second component, and the latent vector V as the input and label. The training is iterative, Fig. 4.5 illustrates a single training round constituting two steps:

(1) The Encoder firstly extracts the latent vector V from X.
(2) We use V to update the EA for one or more epochs.

Hence, V is not constant at each training round, the Encoder should produce a new V from X. The key here is to always use the latest vector V output by the latest Encoder from X. This indirect training mode is used in our framework to tune the EA while maintaining the confidentiality of end devices, as it does not require centralizing sensitive data X on the server.

4.2.3   DeepGuess: Edge-Cloud Fine-Tuning

Noisy latent vectors are the unique information the central server collects from the devices. These vectors are sufficient to tune the Encoder, Decoder, and Main Model using the inductive learning mechanism presented in Sect. 4.2.2. Tuning Encoder's and Decoder's weights is like training a normal Sparse AutoEncoder (SAE). However, this time, the latent vector dimensionality is increased by the Decoder prior to being rebuilt by the Encoder. The SAE training is achieved through the back-propagation of the computed loss function, similar to a standard feed-forward neural network using mini-batch Stochastic Gradient Descent (SGD) [27] algorithm. The choice of the loss function may vary according to the dataset. However, the simple mean square error (MSE) adopted in our experiments works very well in most cases. The Decoder will try somehow to produce a distribution as close as possible to the end-devices data. Accordingly, it is necessary to prevent the SAE from simply copying its input to output during training and performing poorly at the inference stage with the curse of over-fitting. Regularization and sparsity penalties can be used herein. A sparsity penalty is applied to hidden layers to stimulate their activation. The final SAE loss ($MSE + regularizer + sparcity\ penalty$) can be formulated as Eq. (4.2). In the equation, V represents the latent vector of size n and V' is output by EA: $V' = Encoder(Decoder(V))$.

$$L(V, V') = (\frac{1}{n}) \sum_{i=1}^{n} (V - V')^2 + R + \beta \sum_{j=1}^{s_2} KL(p||\hat{p}_j) \qquad (4.2)$$

The regularization term R should result from different regularization techniques (l_0, l_1, l_2 [28], etc.). For the sparsity penalty $\beta \sum_{j=1}^{s_2} KL(p||\hat{p}_j)$, we implemented the sparse activity regularization using Kullback-Leibler (KL) penalty according to [26]. Where $KL(p||\hat{p}_j) = p log \frac{p}{\hat{p}_j} + (1 - p) log \frac{1-p}{1-\hat{p}_j}$ is the KL divergence between a random Bernoulli variable with mean p and a Bernoulli random variable with mean $\hat{p}_j$ [26]. p controls the sparsity level of each layer,s_2 is the number of neurons in the hidden layer, and β is a weight factor. Algorithm 4 summarizes the entire training steps. It illustrates how a differentially private output is generated from the client device and how tuning is conducted on the central server.

The training process, consisting of several iterations, is entirely transparent to the end-users. Whenever a client accesses the service, it only makes inferences by transferring noisy latent encoding to the server and receiving the result regardless of whether the model is trained or not. In order to continuously improve the service quality and client experience, the central server needs data produced by the end devices to train the models. To exchange data in a privacy-conserving manner, the end devices preprocess their sensitive data and only send a noisy latent coding and the corresponding labels to the server. The central server then collects latent coding from multiple devices over a given period, uses it to tune models, and deletes it after tuning. Both SAE and Main model's weights are randomly initialized, and the training round presented in Fig. 4.3, and Algorithm 4 is as follows:

(1) **Encoder Deployment**   The server deploys the latest Encoder to the end devices.
(2) **Features Extraction**   Each device uses the Encoder to perform a forward pass with its local sensitive data and extract important features within a latent vector. The latent vector is further passed to the Noise addition module that incorporates random noise. After that, the noisy latent vector is transmitted to the central server.
(3) **SAE Tuning**   On the server, the SAE is firstly tuned for one or more epochs, while the Main model remains constant. The Decoder takes the noisy latent

features as input and approximates the device's sensitive data. With the generated approximation, the Encoder attempts the inverse operation. The error measured between Decoder input and Encoder output as formulated in Eq. (4.2) is used in the backpropagation process by the optimizer to adjust their weights.

(4) Main Model Tuning Similarly, we keep the SAE constant when training the Main model. The Main model is also tuned for one or more epochs. As shown by Fig. 4.3, the Main model can be designed to take the latent vectors or the Decoder's output as its input. The error between the Main model prediction and the expected ground truth is then used to update its weights. If the SAE is not locked, gradients will flow through the Decoder, making the entire system unstable.

After several rounds, as the SAE improves with training, the Main Model's performance will also improve until the whole system converges. When the system converges, the final Encoder is deployed on the edge devices, where it is only used for inferences until the next training session.

Algorithm 4 Training algorithm

Data:
$S_G \leftarrow$ Encoder's global sensitivity;
$\epsilon \leftarrow$ Privacy budget;
$p \leftarrow$ RUR probability;

1: **for** *round in range* $[0, 1, ..., n_{rounds}]$ **do**   ▷ End clients
2:   **for** *client in client1, client2, ...* **do**
3:     Download the *Encoder* from Server;
4:     Get (x, y), the sensitive data and the target;
5:     $v \leftarrow Encoder(x)$ ▷ Extract features
6:     $v \leftarrow$ Add noise (Eq. (4.5));
7:     Offload v and target y to the Server;
8:   **end for**   ▷ Server
9:   **for** *client in client1, client2, ...* **do**   ▷ Collect latent vectors from clients
10:     Get (v_i, y_i);
11:   **end for**   ▷ Tuning
12:   $(V, Y) \leftarrow$ All latent vectors and targets;   ▷ Tune the EA
13:   **for** epochs in $[1, 2, ..., n_{epochs}]$ **do**
14:     $X' \leftarrow Decoder(V)$ ▷ Estimate extracted utility from X
15:     $V' \leftarrow Encoder(X')$
16:     Compute the error between V, V' (Eq. ((4.2)));
17:     Optimize *Encoder* and *Decoder* through back-propagation;
18:   **end for**   ▷ Tune the main model
19:   **for** *epochs in* $[1, 2, ..., n_{epochs}]$ **do**
20:     $Y' \leftarrow Model(V)$;
21:     Compute the error between Y, Y';
22:     Optimize *Model* through back-propagation;
23:   **end for**
24: **end for**

4.2.4 Enhancing Privacy with Noise

Only Extracting and transferring features to the server does not guarantee total privacy since it has been proven [29, 30] that it is possible to leak sensitive information with these features. The trained model is capable of memorizing some sensitive information, as shown by Fig. 4.7, and may be used by a malicious adversary later to leak the memorized knowledge, thus violating privacy. In addition, some sensitive information from the input can remain perceptible in the latent feature vector. To reinforce privacy by making it harder to disclose sensitive features, we add random noise to the latent vector. We apply two levels of randomization to the latent features before transferring them to the central server. The first level, called Randomized Units Response (RUR), is inspired by the randomized response mechanism presented in heading "Randomized Response: Coin Flips". We use the Laplace DP additive noise mechanism [31] for the second level of randomization.

Client data may include images, text, audio, etc. Adding noise directly to the data will involve designing a specific DP mechanism for each data type. It is important to note that DP is not a catch-all solution for any task or dataset but a methodology for enforcing privacy in a framework. For example, only categorical data and strings are supported by Google's RAPPOR [24] DP framework. Nevertheless, our solution does not make any assumptions about the dataset. We apply noise to the latent vector rather than the input data. During the client side's features extraction, from the lowest to highest Encoder's layers, the input dimensionality will be gradually reduced until the units on the last layer produce the final output, which can be a 1-D latent vector. Here, the Encoder is our query function f and maps the user's input to a latent representation containing activations of its last layer.

In a neural network, the output of each layer $h_{W,b}(X) = \phi(XW + b)$ is used as input for the next layer. X is what the layer receives in input, W is a weights matrix, b is a bias vector, ϕ is called the activation function. As we can observe that $XW + b$ is a linear operation, the role of ϕ is to incorporate nonlinearity and help the network learn much more complex structures from the data. For reasons we would explain later, all Encoder's layers could use any available activation function except for the last layer, whose activation had to be bounded within a certain range. Then we have $\alpha \leq f_u \leq \beta$ with α and β as the minimum and maximum activation for a unit u on the last layer. Possible bounded activation functions [32] may include the Hyperbolic Tangent ($f(x) \in [-1, 1]$), the Binary Step function ($f(x) \in \{0, 1\}$), the Sigmoid function ($f(x) \in [0, 1]$) and the bounded variant of the Rectified Linear Unit (ReLU) proposed in [33].

1. **Randomized Units Response (RUR):** For DP to use Randomized Response, it only works when the response is a binary attribute (YES or NO) [31] or a categorical value [24]. However, in our solution, the latent vector contains real-value numbers. We could binarize the latent vector as suggested by Pathum et al. [34] with an algorithm called LATENT. Nevertheless, the problem with

LATENT is that the algorithm was quite complex by itself, and it would require heavy computation from the client device.

$$RUR(f_u, p) = \begin{cases} \mathscr{R}, \ \mathscr{R} \in [\alpha, \beta]; & prob: \ p \\ f_u & ; & prob: \ (1-p)) \end{cases} \quad (4.3)$$

With RUR, we have taken a more straightforward approach. Equation (4.3) illustrates the RUR mechanism performed by randomizing each unit activation value presented in the latent vector. The value is replaced by the random number $\mathscr{R}$ drawn from the range $[\alpha, \beta]$ with a probability of p and is preserved with a probability of $1 - p$. This is the first reason we need a bounded activation for the last layer. It ensures that an adversary cannot easily find patterns to recognize True activation values from random values within the latent vector. For $p = 0$ (utility preserved and low privacy), all True activations are preserved. For $p = 1$ (No utility and perfect privacy), all activations are replaced by random values. Since p is defined by the client, RUR provides it with certain refutability to the data it transfers to the cloud. As we will see in the experiments section, RUR also acts as a regularizer by enforcing the network to learn the overall data distribution instead of a particular client distribution. It also helps overcome over-fitting during training.

2. **Additive noise:** Some True activations can pass through the RUR. To further enforce uncertainty, noise is added to each variable using the Laplace-DP mechanism. We first define the data sensitivity according to a single unit response base on the global sensitivity given in Eq. (4.1). Since the last layer's activation is bounded by α and β, the sensitivity of f for each unit can be defined as:

$$S_{f_u} = \beta - \alpha \quad (4.4)$$

If we consider that the last layer has k neural units, the overall sensitivity will be $S_f = k(\beta - \alpha)$.

$$f'_u = RUR(f_u) + \mathscr{R}, \mathscr{R} \in Lap(\frac{S_f}{\epsilon}) \quad (4.5)$$

At the end, each value f_u in the latent vector is replaced by f'_u defined in Eq. (4.5). This time the random number $\mathscr{R}$ is drawn from the Laplace distribution $Lap(\frac{S_f}{\epsilon})$ and the final result f' is sent to the central server for further processing. The client could control his total privacy budget with p and ϵ. For stronger privacy, he could increase the random response likelihood by setting a high p value and/or a high noise by using a smaller ϵ value. As a side effect, setting high privacy budgets could deteriorate the quality of the service he received. Then he has to find a good balance between privacy and the required service quality. Combining RUR with the Laplace-DP mechanism helps us enable $\epsilon - LDP$ using the randomized response mechanism on real-value responses instead of binary responses. Our approach is much more resource-efficient and quicker to compute than the LATENT [34]. This is crucial if we want to reduce the processing and latency on the client side.

4.3 Framework Evaluation

The proposed learning technique in this chapter was evaluated using three different datasets. We began with the basic MNIST [35] and then progressed to more challenging datasets such as the CIFAR10 [36] and the UCI-HAR [37]. These datasets offer various degrees of difficulty and are frequently used to evaluate the robustness of learning frameworks. In our experiments, we first demonstrate that the inductive learning technique presented in Sect. 4.2.2 can converge by comparing our framework to a Centralized-ML strategy that employs traditional mini-batch stochastic gradient descent without any privacy-preserving mechanism. We then investigated the impact of various privacy budgets on the performance of our system and compared our experimental results to those of existing SOTA frameworks. Our simulation environment has one server and four edge devices. The edge devices are emulated using four Raspberry Pis. During training, we split each dataset equally among the four Raspberry Pis. We use Python Keras API [38], a high-level API of Google's TensorFlow engine [39] for the experiments.

4.3.1 Datasets

The MNIST [35] is about handwritten digits. Each image contains a single digit in the grayscale format of size 28x28. There are ten digits from 0 to 9 (10 classes) with 60,000 training samples and 10,000 testing samples. The CIFAR-10 dataset consists of 60,000 32×32 color images in 10 classes. There are 50,000 training images and 10,000 testing images. We resized the MNIST dataset to 32×32 in order to have the same width and height for both MNIST and CIFAR.

The UCI-HAR dataset is a Human Activity Recognition (HAR) database built from the recordings of 30 subjects performing activities of daily living while carrying a waist-mounted smartphone with embedded inertial sensors. Each subject performed six activities (Walking, Walking_Upstairs, Walking_Downstairs, Sitting, Standing, Laying). The phone's embedded sensors (accelerometer and gyroscope) record data about each activity. After preprocessing the raw database, each input instance has 9 channels, each with a length of 128. With 7352 training instances and 2947 testing instances, the dataset end-up with a shape of (7352,128,9) for the training set and (2947,128,9) for the testing set.

4.3.2 Models and Hyper-Parameters

The Encoder, Decoder, and Main model for each dataset are designed as shown in Table 4.1. Our experiment considers the configuration of Fig. 4.3, with Main Model 1 as our task model. All the hyperparameters used are empirically inspired by existing networks commonly used to solve the used datasets.

MNIST&CIFAR The CIFAR and MNIST networks have the same design. The Encoder and Decoder are designed following the symmetric AutoEncoder paradigm. The last value of the 'output shape' represents the number of output units of the corresponding layer. The default strides (strides = 1) and padding = "same" are used for the convolution (Conv) layers. The MaxPooling and UpSampling layers are 2*2. The final FC (fully connected) layer of the Main model employs a Softmax

Table 4.1 Network structure per dataset. Although the Encoder and Decoder are symmetric, they could be designed differently

	MNIST/CIFAR		UCI-HAR	
	Layer	Output shape	Layer	Output shape
Encoder	input	32*32*1/3	input	128*9
	conv2d	32*32*256	conv1d	128*512
	max_pool	16*16*256	max_pool	64*512
	conv2d	16*16*128	conv1d	64*256
	max_pool	8*8*128	max_pool	32*256
	conv2d	8*8*32	conv1d	32*128
	–	–	max_pool	16*128
	–	–	conv1d	16*64
	–	–	max_pool	8*64
Decoder	input	8*8*32	input	8*64
	conv2d	8*8*32	conv1d	8*64
	up_sample	16*16*32	up_sample	16*128
	conv2d	16*16*128	conv1d	16*128
	up_sample	32*32*128	up_sample	32*128
	conv2d	32*32*256	conv1d	32*256
	conv2d	32*32*1/3	up_sample	64*256
	–	–	conv1d	64*512
	–	–	up_sample	128*512
	–	–	conv1d	128*9
Main model	input	8*8*32	input	8*64
	flatten	2048	flatten	512
	fc	256	fc	128
	fc	128	fc	64
	fc	64	fc	6
	fc	10	–	–

activation to output the probability of belonging to a class. The encoder's last layer uses Tanh activation when training with the noise module. As a recall, when using Laplace noise, we need bounded activations to estimate the global sensitivity. Finally, the images are normalized by dividing by 255.

UCI-HAR Since this dataset does not include images, the Encoder, Decoder, and Main model configurations differ. On the other hand, it was proved in [40] that convolutional networks could perfectly solve the classification task associated with this dataset. Therefore, the Encoder and Decoder are designed using the 1D versions of Conv, MaxPooling, and UpSampling layers. The strides, padding, and activations are specified similarly to the case of MNIST&CIFAR.

Centralized-ML Network The Centralized-ML model is obtained by stacking the Main model on top of the Encoder. It does not include any DP mechanisms. It uses the same learning rate, regularizers, and decay step as our system. The training of the Centralized-ML is centralized on the server.

For each test, we train our system for 60 rounds and the Centralized-ML network for 60 epochs. We use a mini-batch size of 32, the adam optimizer, a mean square loss for the AutoEncoder, and a cross-entropy loss for the Centralized-ML and the Main model. The learning rate starts from 0.001 with a decaying polynomial step of 10,000. Since our goal is not to outperform the state-of-the-art results for each dataset, the hyperparameters are not specially fine-tuned. First, we compared the Centralized-ML model to our system without the noise module. Secondly, we incorporate the noise module using various privacy budgets and analyze the effect on the client's privacy and the system's accuracy.

4.3.3   Comparison Results with the Centralized-ML

Results Without Differential Privacy We initially removed the noise module and compared our framework to the Centralized-ML model to validate the intuition behind the inductive learning process. Even without the noise module, our system ensures some level of privacy since the sensitive data stays on end devices. Figure 4.6 plots the obtained results on the testing datasets. The first observation from this Figure is the slow convergence of DeepGuess compared with the Centralized-ML, typically on the MNIST and CIFAR datasets. A possible explanation for this slow convergence is that, for our Main model to improve, the AutoEncoder has to reduce its losses first. For MINST, the slow convergence phase is visible in the first 10 rounds. We obtain 92% testing accuracy, while the Centralized-ML model is already at 96%. The learning speed on the CIFAR and UCI-HAR datasets is stable but not as fast as the Centralized-ML. Table 4.2 shows the best result obtained for each dataset. Although DeepGuess did not outperform Centralized-ML in testing accuracy, the obtained result was still satisfying.

These results confirm the motivation behind inductive learning and prove that learning from a dataset without involving it in any back-propagation process is

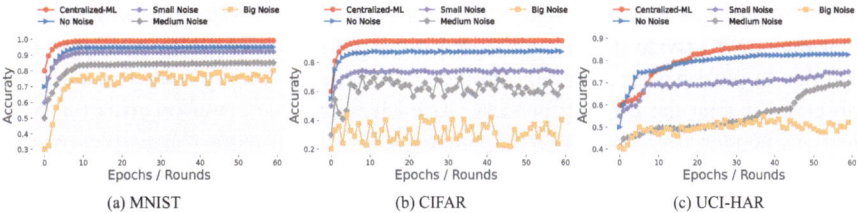

(a) MNIST   (b) CIFAR   (c) UCI-HAR

Fig. 4.6 The Centralized-ML and DeepGuess testing results with and without DP noise. Without the noise module, our framework can achieve satisfying results on the three datasets. The higher the noise, the more difficult the training is. (**a**) MNIST. (**b**) CIFAR. (**c**) UCI-HAR

Table 4.2 Final training accuracy

Framework	Privacy budget	MNIST	CIFAR	UCI-HAR
Centralized-ML	N/A	0.99	0.94	0.89
DeepGuess: No Noise	N/A	0.95	0.89	0.83
DeepGuess: Small Noise	$(p = 0.1, \epsilon = 30)$	0.90	0.79	0.75
DeepGuess: Medium Noise	$(p = 0.2, \epsilon = 20)$	0.85	0.61	0.70
DeepGuess: Big Noise	$(p = 0.3, \epsilon = 10)$	0.79	0.42	0.55

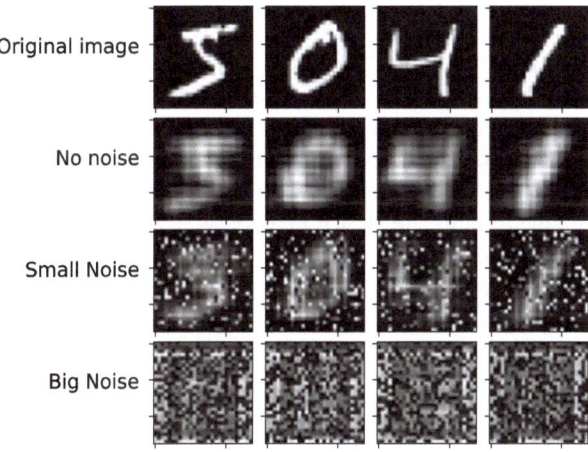

Fig. 4.7 Decoder's output from some latent vectors received by the server

possible. Our system's performance is nearly identical to the Centralized-ML on all three datasets. We then train a centralized AI model without moving the sensitive data to the cloud and without back-propagation on end devices. Figure 4.7 shows the Decoder's output for some latent vectors received by the server for the MNIST dataset. Although the server never had access to the sensitive end devices' data, the Decoder can accurately estimate their distribution. The reconstructed digits from the latent vector seem slightly bigger and blurry, but they are clearly identifiable. Therefore, inductive learning without the noise module cannot guarantee strong confidentiality to end-users, particularly for image inputs. It does, however, assure that their raw private data is not stored on a central server while providing data utilities to third parties in the meantime. The main weakness of our framework is observed during the training process. At each round, the client receives the most recent Encoder from the server. For many steps, it will extract and sends back a latent vector. The training is a bit expensive, but this cost is largely compensated during inferences. At inference time, the client gets the most recent Encoder from the server. It only offloads the latent vector to the server. Alternatively, if the client's device is powerful enough, the entire inference process could be performed on the device, further reducing the network cost.

Results with Differential Privacy The DeepGuess framework can be tuned in a more privacy-preserving manner. The noise module allows the client to degrade the latent vector by adding $epsilon - LDP$ noise. This could prevent the cloud server from fully disclosing sensitive information. Furthermore, the latent vector will appear noisy to the attacker if intercepted. Recall that the noise is controlled by two parameters: p and ϵ. A high value of p will include more randomization within the latent vector by substituting some units' responses with random values. In contrast, a small value of ϵ will increase the Laplace noise added to them. So high noise means a high value of p or a small value of ϵ.

Thus, we started by defining three levels of privacy budgets starting from small noise $(30, 0.1)$, medium noise $(20, 0.2)$, and big noise $(10, 0.3)$. Figure 4.6 also shows the testing results per privacy budgets and dataset. Increased privacy comes at the expense of decreased accuracy. On the MNIST task, our system could yield good results even with big noise. However, learning becomes unstable as noise increases for more complex tasks, such as CIFAR and UCI-HAR. The best testing results obtained during the training are shown in Table 4.2. We can see from Fig. 4.7 that there is no strong privacy without noise since the sensitive data is recoverable. With small noise, the recovered digits look damaged. However, some utilities are well preserved. When we increase the noise, we gain more privacy, but the image loses its utility.

Preventing Leakage with Dynamic Privacy Budget Higher noise provides better privacy but tends to sacrifice all the data's usefulness. Existing privacy-preserving systems will generally define a unique privacy budget for all clients. However, it is difficult to know when and how the client is willing to set his own privacy budget. Our system can assist the end-user in finding a balance between his privacy and the utilities he is willing to provide to the service provider. Different instances of the client's sensitive data may necessitate varying amounts of noise. Instead of arbitrarily establishing a constant privacy budget, the Decoder is also forwarded to the client's device to evaluate the reconstruction risk by experimenting with various privacy budgets. The client might examine the output of the Decoder and become aware of potential reconstructions from the latent vector on the cloud. He then adjusts his budget based on the utilities he wants to reveal or hide. Without this feature, most clients will tend to set high noise, which may not be suitable for the services offered.

4.3.4 Comparison Against Existing Frameworks

We compare our framework results to three other existing differentially private deep learning mechanisms, which are:

- **LATENT [34]:** The LATENT framework splits the neural network into the feature extraction module, the randomization module, and the fully connected module. The client runs the pre-trained feature extraction module on its private dataset, followed by the randomization module, which adds DP noise to the extracted feature before transferring it to the server. The randomization module in LATENT converts each value of the extracted features into binary and applies the randomized response mechanism at a binary level. Afterward, the randomized binary representation is transferred to the server and used for training and inferencing on the fully connected module.
- **Arden [6]:** The Arden framework also partitions the DNN across mobile devices and cloud data centers. A simple data transformation is performed on the mobile device, while the resource-hungry training and inference rely on the cloud data center. To protect sensitive information, a lightweight privacy-preserving mechanism consisting of arbitrary data nullification and random noise addition is introduced.
- **Osia et al. [9]:** In this work, the authors propose a framework that lets the IoT device run the initial layers of the neural network and then send the output to the cloud to feed the remaining layers and produce the final result. In order to ensure users' privacy, the authors introduce Siamese fine-tuning to significantly reduce the level of unnecessary and potentially sensitive information in personal data.

As shown in Fig. 4.8 and Table 4.3, we defined three levels of noise for each framework (Small, Medium, Big) and displayed the accuracy of each strategy on our three datasets(MNIST, CIFAR, UCI-HAR). We observe that even with high noise applied, the accuracy of our framework is comparable to, or sometimes better than, that of other strategies. Further investigation has revealed that other strategies (LATENT, Arden, Osia et al.) require the client side feature extraction module to

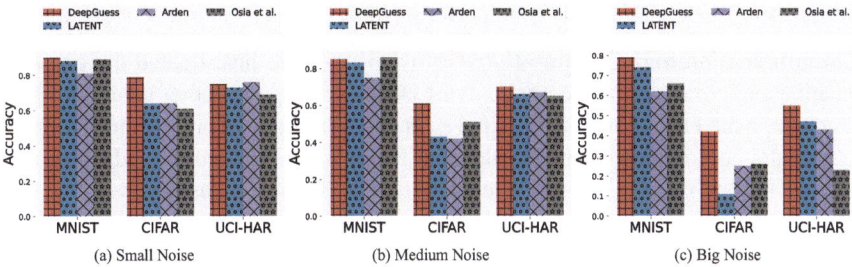

Fig. 4.8 DeepGuess achieves comparable or even higher accuracy than other solutions. (**a**) Small noise. (**b**) Medium noise. (**c**) Big noise

Table 4.3 Final accuracy comparison with existing frameworks

		Dataset		
Strategy	Noise	MNIST	CIFAR	UCI-HAR
Centralized-ML	*N/A*	0.99	0.94	0.89
DeepGuess	*N/A*	**0.95**	**0.89**	**0.83**
	Small	**0.90**	**0.79**	0.75
	Medium	0.85	**0.61**	**0.70**
	Big	**0.79**	**0.42**	**0.55**
Latent [34]	*N/A*	**0.95**	0.79	0.82
	Small	0.88	0.64	0.73
	Medium	0.83	0.43	0.66
	Big	0.74	0.11	0.47
Arden [6]	*N/A*	0.94	0.68	**0.83**
	Small	0.81	0.64	**0.76**
	Medium	0.75	0.42	0.67
	Big	0.62	0.25	0.43
Osia et al. [9]	*N/A*	**0.95**	0.67	0.81
	Small	0.89	0.61	0.69
	Medium	**0.86**	0.51	0.65
	Big	0.66	0.26	0.23

The bold values in Table 4.3 represent the highest accuracy achieved for each dataset and noise level combination when comparing across all strategies (Centralized-ML, DeepGuess, Latent, Arden, and Osia et al.), indicating the superior performance of the most effective method under specific conditions. For instance, at the 'Small' noise level with the MNIST dataset, the accuracy values are 0.90, 0.88, 0.81, and 0.89 for DeepGuess, Latent, Arden, and Osia et al., respectively. Here, our method, DeepGuess, achieves the highest accuracy (0.90), which is highlighted in bold, emphasizing the superior performance of our approach under the "small" noise conditions

be pre-trained on an auxiliary dataset before they can work. Consequently, only the prediction module deployed in the cloud server is actually trained on the sensitive dataset. We use the tiny ImageNet dataset as the auxiliary dataset for those strategies. Since our framework does not require pretraining, both the features extraction module and the prediction module for our strategy are trained directly on the dataset. This explains why it outperforms other solutions in terms of accuracy.

Communication and Computational Burden We also investigated the communication and computational complexity of DeepGuess against other solutions. This is accomplished by measuring the time each solution spent on pre-training, computation, and communication under identical hardware and network conditions. The average time of each solution on the experimental datasets is displayed in Fig. 4.9. The Centralized-ML strategy has a lower computational complexity because it has no privacy-preserving mechanisms. Our strategy and Arden follow the Centralized-ML since they have similar computational and communication complexity. The computational complexity of the LATENT framework is relatively high due to its randomization module, which appears to be computationally expensive. This

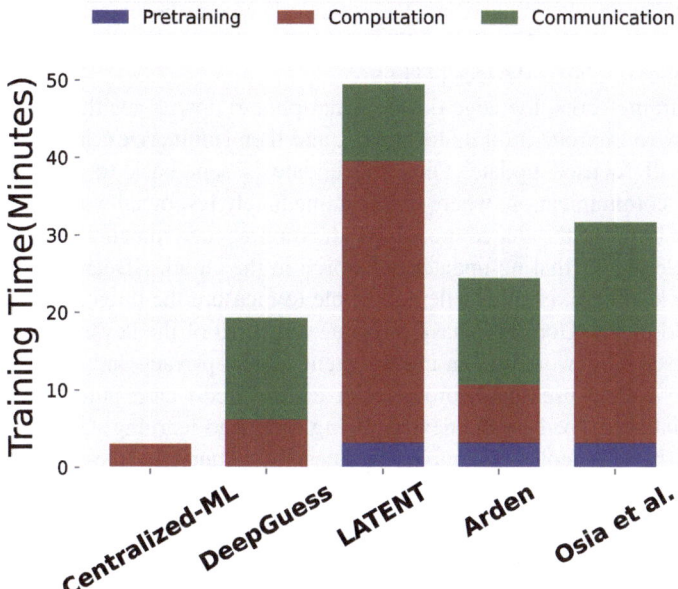

Fig. 4.9 The LATENT's randomization module is computationally expensive compared to other solutions. DeepGuess eliminates the need for pre-training and trains faster

is because the randomization module converts each value in the latent vector into binary before applying the randomized response strategy at the bit level. Additionally, the training time of our framework is further reduced since we are not required to pre-train the feature extractor as with other techniques. Therefore, we can conclude that DeepGuess provides the best trade-off regarding Computation, Communication, Accuracy, and Privacy.

4.4   Related Work

This section discusses earlier work on privacy-preserving frameworks for the edge intelligence environment. Three main approaches stand out from the literature. *Centralized approach:* This is the conventional cloud computing approach dedicating all processing to a central server. *Decentralized approach:* all processing is performed at the edge of the network. It can also take the form of a collaboration between end-devices. *Hybrid approach:* It is neither centralized nor decentralized, the processing is shared between the devices and the central server.

The centralization and processing of device data could lead to privacy issues by exposing some confidential information. To avoid these problems in its Android apps, Google had developed Federated Learning (FL) [41], a distributed learning system. Federated learning is an AI framework where the objective is to train a high-quality centralized model while the training data remains decentralized on a large number of clients devices [41].

In FL frameworks, the edge device (smartphone) downloads the central model, learns how to improve it on its local data, and then summarizes the improvements like a small focused update. Only the update is sent back to the cloud using encrypted communication, where it gets immediately leveraged with other devices' updates to improve the shared model. All the training data remains on edge devices, and Google claims that no updates are stored in the cloud. GBoard, the Android's predictive keyboard, is an excellent example to measure the threat. Centralizing its data would enable Google to have direct access to all of the keystrokes performed by each user. This would be an infringement of user privacy and could also allow Google to collect user passwords, secret codes, credit card numbers, and other confidential text typed on the device. Using federated learning, Google addresses this issue by only collecting model updates. Unfortunately, the FL is still at its infancy stage and faces critical challenges that remain unresolved.

- **Network cost [42]:** In FL, the trade-off between privacy and other factors such as communication costs is not well balanced. Federated networks are potentially made up of large number of smart devices which may lead to a heavy network activity during each training iteration.
- **Devices diversity [42]:** Due to the hardware diversity of each device, federated networks could contain different types of devices with unbalanced resource capabilities. For this reason, only eligible devices may participate in the training. Moreover, in order to reduce the communication and power bills, the selected devices must be plugged into a power source and connected to a wifi network.
- **Concerns on privacy [42, 43]:** The FL has not yet fully achieved its main privacy objective. Training the model at the data source, then sending model updates (gradient information) to the server, rather than raw user data, might not guarantee a total privacy, as studies show that an interceptor could disclose sensitive information with these updates.

Faced with the difficulties encountered by decentralized training strategies, a Hybrid option preserves the conventional centralized approach, except that the data is pre-processed and anonymized at the device level before being transferred to the central server. Such a strategy was adopted by the hybrid framework proposed by Osia et al. [9]. Instead of running the entire process on the server, their system breaks down the DNN into a feature extraction module, which should be deployed on the client's device, and a classification module that operates in the cloud. The motivation is to let the edge device run the initial layers of the neural network and then send the output to the cloud to feed the remaining layers and generate the final output. The service provider should have pre-trained the feature extractor on a public dataset before releasing it to the devices. Various techniques can be applied to the feature extractor's output for better privacy. Firstly a dimensionality reduction technique through PCA (Principal component analysis) [44] is to be used to reduce the feature dimensionality. Secondly, a technique referred to as Siamese fine-tuning has helped refine the feature extractor so that features of the same class fall within a small neighborhood of each other. Finally, noise addition is used to increase the inference uncertainty of unauthorized tasks. A similar process is adopted in most of proposed hybrid frameworks [6–9]. In summary, privacy-preserving in edge computing is a hot topic, and several challenges are still present. Our framework proposed in this chapter mainly addresses the issue of the unnecessary reliance on a public dataset or pre-train model to build the features extractor that has never been addressed before.

4.5 Summary

We introduced inductive learning, a new mechanism for training a feature extractor and a centralized neural network while keeping sensitive data at the network's edge. We also suggested a new LDP algorithm combining randomized response and Laplace-DP mechanisms for better privacy. Both combined allowed training a centralized DNN with the end devices as the only data source in a privacy-preserving manner. Previous works typically use public datasets or publicly available models to construct a feature extractor. Our work proves that an efficient feature extractor can be built without a public dataset or pre-trained model while preserving end-user privacy. We also consider the fact that the end devices have constrained resources by moving heavy processing to the server. Our work opens several future directions for research. In the next stage, we intend to investigate strategies to reduce training overheads and explore new applications of inductive learning.

References

1. Zanella, A., Bui, N., et al.: Internet of things for smart cities. IEEE Internet Things J. **1**(1), 22–32 (2014)
2. Dwork, C.: Differential privacy: a survey of results. In: International Conference on Theory and Applications of Models of Computation, ser. Lecture Notes in Computer Science, vol. 4978, pp. 1–19. Springer (2008)
3. Narayanan, A., Shmatikov, V.: Robust de-anonymization of large sparse datasets: a decade later. Princeton Univ. **21**, 1–10 (2019)
4. Zhuang, F., Qi, Z., et al.: A comprehensive survey on transfer learning. Proc. IEEE **109**(1), 43–76 (2021)
5. Tan, C., Sun, F., et al.: A survey on deep transfer learning. In: Artificial Neural Networks and Machine Learning - ICANN 2018, pp. 270–279. Springer International Publishing (2018)
6. Wang, J., Zhang, J., et al.: Not just privacy: Improving performance of private deep learning in mobile cloud. In: Proceedings of the 24th ACM SIGKDD International Conference on Knowledge Discovery & Data Mining, pp. 2407–2416. ACM (2018)
7. Mao, Y., Yi, S., et al.: A privacy-preserving deep learning approach for face recognition with edge computing. In: Proc. USENIX Workshop Hot Topics Edge Comput. (HotEdge), pp. 1–6. USENIX Association (2018)
8. Ghosh, A.M., Grolinger, K.: Deep learning: Edge-cloud data analytics for IoT. In: 2019 IEEE Canadian Conference of Electrical and Computer Engineering, CCECE 2019, Edmonton, AB, Canada, May 5–8, 2019, pp. 1–7. IEEE (2019)

 9. Osia, S.A., Shamsabadi, A.S., et al.: A hybrid deep learning architecture for privacy-preserving mobile analytics. IEEE Internet Things J. **7**(5), 4505–4518 (2020)
10. McMahan, B., Moore, E., et al.: Communication-efficient learning of deep networks from decentralized data. In: Singh, A., Zhu, X.J. (eds.) Proceedings of the 20th International Conference on Artificial Intelligence and Statistics, AISTATS 2017, 20–22 April 2017, Fort Lauderdale, FL, USA, ser. Proceedings of Machine Learning Research, vol. 54, pp. 1273–1282. PMLR (2017)
11. Qiu, G., Gui, X., Zhao, Y.: Privacy-preserving linear regression on distributed data by homomorphic encryption and data masking. IEEE Access **8**, 107601–107 613 (2020)
12. LeFevre, K., DeWitt, D.J., Ramakrishnan, R.: Mondrian multidimensional k-anonymity. In: 22nd International Conference on Data Engineering (ICDE'06), pp. 25–36. IEEE (2006)
13. Christ, M., Radway, S., Bellovin, S.M.: Differential privacy and swapping: examining de-identification's impact on minority representation and privacy preservation in the U.S. census. In: 43rd IEEE Symposium on Security and Privacy, SP 2022, San Francisco, CA, USA, May 22–26, 2022, pp. 457–472. IEEE (2022)
14. Hafner, K.: If you liked the movie, a Netflix contest may reward you handsomely. New York Times **2** (2006)
15. Narayanan, A., Shmatikov, V.: Robust de-anonymization of large sparse datasets. In: 2008 IEEE Symposium on Security and Privacy (SP 2008), pp. 111–125. IEEE Computer Society (2008)
16. Maas, A.L., Daly, R.E., et al.: Learning word vectors for sentiment analysis. In: Proceedings of the 49th Annual Meeting of the Association for Computational Linguistics: Human Language Technologies. Association for Computational Linguistics, Portland, Oregon, USA, June 2011, pp. 142–150 (2011)
17. Kim, J.: A method for limiting disclosure in microdata based on random noise and transformation. In: Proceedings of the Section on Survey Research Methods, pp. 303–308. American Statistical Association Alexandria, VA (1986)
18. Kim, J., Winkler, W.: Multiplicative noise for masking continuous data. Statistics **1**, 9 (2003)
19. Dwork, C., Roth, A., et al.: The algorithmic foundations of differential privacy. Found. Trends Theor. Comput. Sci. **9**(3–4), 211–407 (2014)
20. Nissim, K., Steinke, T., et al.: Differential privacy: A primer for a nontechnical audience. In: Privacy Law Scholars Conf., vol. 21, p. 209. HeinOnline (2017)
21. Kasiviswanathan, S.P., Lee, H.K., et al.: What can we learn privately? SIAM J. Comput. **40**(3), 793–826 (2011)
22. Warner, S.L.: Randomized response: A survey technique for eliminating evasive answer bias. J. Am. Stat. Assoc. **60**(309), 63–69 (1965)
23. Kairouz, P., Oh, S., Viswanath, P.: Extremal mechanisms for local differential privacy. In: Advances in Neural Information Processing Systems (NeurIPS 2014), vol. 27, pp. 2879–2887. Curran Associates (2014)
24. Erlingsson, Ú., Pihur, V., Korolova, A.: RAPPOR: Randomized aggregatable privacy-preserving ordinal response. In: Proceedings of the 2014 ACM SIGSAC Conference on Computer and Communications Security (CCS 2)14), pp. 1054–1067. ACM (2014)
25. Zhou, Z., Chen, X., et al.: Edge intelligence: Paving the last mile of artificial intelligence with edge computing. Proc. IEEE **107**(8), 1738–1762 (2019)
26. Ng, A., et al.: Sparse autoencoder. Stanford Univ. **72**(2011), 1–19 (2011)
27. Bottou, L.: Stochastic gradient descent tricks. In: Montavon, G., Orr, G.B., Müller, K. (eds.) Neural Networks: Tricks of the Trade - Second Edition, ser. Lecture Notes in Computer Science, vol. 7700, pp. 421–436. Springer (2012)
28. Tibshirani, R., Wasserman, L.: A closer look at sparse regression. Lecture Notes (2016)
29. Song, C., Ristenpart, T., Shmatikov, V.: Machine learning models that remember too much. In: Proceedings of the 2017 ACM SIGSAC Conference on Computer and Communications Security, pp. 587–601 (2017)
30. Fredrikson, M., Jha, S., Ristenpart, T.: Model inversion attacks that exploit confidence information and basic countermeasures. In: Proceedings of the 22nd ACM SIGSAC Conference on Computer and Communications Security, pp. 1322–1333 (2015)

31. Wang, Y., Wu, X., Hu, D.: Using randomized response for differential privacy preserving data collection. In: Proceedings of the Workshops of the EDBT/ICDT 2016 Joint Conference, EDBT/ICDT Workshops 2016, Bordeaux, France, March 15, 2016, ser. CEUR Workshop Proceedings, vol. 1558. CEUR-WS.org (2016)

32. Nwankpa, C., Ijomah, W., et al.: Activation functions: Comparison of trends in practice and research for deep learning. Preprint. arXiv:1811.03378 (2018)

33. Liew, S.S., Khalil-Hani, M., Bakhteri, R.: Bounded activation functions for enhanced training stability of deep neural networks on visual pattern recognition problems. Neurocomputing **216**, 718 734 (2016)

34. Arachchige, P.C.M., Bertok, P., et al.: Local differential privacy for deep learning. IEEE Internet Things J. **7**(7), 5827–5842 (2020)

35. Deng, L.: The mnist database of handwritten digit images for machine learning research [best of the web]. IEEE Signal Process. Mag. **29**(6), 141–142 (2012)

36. Krizhevsky, A., Hinton, G.: Learning multiple layers of features from tiny images. Technical Report, University of Toronto (2009)

37. Garcia-Gonzalez, D., Rivero, D., et al.: A public domain dataset for real-life human activity recognition using smartphone sensors. Sensors **20**(8), 2200 (2020)

38. Chollet, F., et al.: Keras. https://github.com/fchollet/keras (2015)

39. Abadi, M., Agarwal, A., Barham, P., Brevdo, E., Chen, Z., Citro, C., Corrado, G.S., Davis, A., Dean, J., Devin, M., et al.: TensorFlow: large-scale machine learning on heterogeneous systems. Available at: http://tensorflow.org/ (2015)

40. Dua, N., Singh, S.N., Semwal, V.B.: Multi-input CNN-GRU based human activity recognition using wearable sensors. Computing **103**(7), 1461–1478 (2021)

41. Konečný, J., McMahan, H.B., et al.: Federated learning: Strategies for improving communication efficiency. Preprint. arXiv:1610.05492, vol. abs/1610.05492 (2016)

42. Li, T., Sahu, A.K., et al.: Federated learning: Challenges, methods, and future directions. IEEE Signal Process. Mag. **37**(3), 50–60 (2020)

43. Kairouz, P., McMahan, H.B., et al.: Advances and open problems in federated learning. Found. Trends Mach. Learn. **14**(1–2), 1–210 (2021)

44. Kim, T., Choe, Y.: Fast circulant tensor power method for high-order principal component analysis. IEEE Access **9**, 62478–62492 (2021)

Chapter 5
SDN-Based and Energy Aware Offloading

Abstract The chapter begins by analyzing the energy consumption of devices, illuminating the causes of power drain, and emphasizing the need for optimized solutions. It then examines task execution, contrasting local execution's benefits with offloading's, and provides information on when and why offloading may be the better option. The chapter also explores the core of offloading optimization and offers methods to improve energy efficiency. It makes a distinction between offloading for a single device on battery power and offloading for many devices on battery, and it provides specific advice for each situation. By the end of this chapter, readers will have a thorough awareness of the difficulties involved in planning and putting into practice offloading methods. They will also have the practical knowledge necessary to make wise decisions.

Keywords Energy consumption · Energy efficiency · Power drain · Task execution · Offloading · Optimization · Energy efficiency · Local execution · Practical knowledge

5.1 Introduction

The IoV as a typical application scenario of the Internet of Things (IoT) has been widely investigated in recent years. It's widely considered as the most promising paradigm for the future intelligent transportation system (ITS) by integrating cutting-edge technologies such as AI, 5G, Big Data, etc. Not only vehicles will be connected in IoV, but also various types of components like roadside units (RSU), mobile devices from passengers and pedestrians will also join the IoV, which will make the IoV a very complicated network that needs real-time and dynamic wireless connection and high-speed connection to the core network.

The 5G network is expected to be the backbone of IoV for its high bandwidth and important features like Device-to-Device (D2D) communication and massive Multiple Input Multiple Output (MIMO), which are leveraged to provide massive edge device collaboration without using the core network. Integrating the SDN with fog computing is considered as the most promising architecture to address

© The Author(s), under exclusive license to Springer Nature Singapore Pte Ltd. 2024
Y. Zhai et al., *Edge Computing Resilience*, SpringerBriefs in Computer Science,
https://doi.org/10.1007/978-981-97-6998-8_5

the processing latency, dynamic connection, and inefficient network management problems in IoV systems [1–4]. The SDN controller is the centralized control module in an SDN control domain, which will collect network information such as the network statistics, network load measurements, vehicle location and mobility, resource utilization, supported application types. With the gathered network information, the SDN controller could interact with vehicles, RSUs, Fog or Cloud through the OpenFlow protocol to send flow table to the network or to set data forwarding rules to control the data flow. In the IoV, many devices like onboard cameras and embedded sensors continuously collect a large amount of data from the vehicles and environment. After handling and analyzing the data, the results generated are transferred back to the actuators like the control system in the vehicle and transportation control units in real-time [5]. Collaborating with cloud computing can significantly increase the computing capability for IoV devices, when a large number of edge devices request computing resources from the cloud center simultaneously, it may cause data transmission congestion, thus prolonging the execution time of the task [6–9]. Fog computing, as a new computing paradigm, could perfectly be integrated with SDN based IoV [2, 10]. With fog computing, the application can benefit from low latency and reduce the amount of data traffic required to be sent to the backhaul network [11].

IoV still faces many challenges related to limited computational resources and capability of on-board equipment, whereas many new types of applications such as AI based environment detection and personalized navigation emerge in IoV are response-sensitive which demand complex computation and real-time analysis. SDN and fog computing based IoV allow vehicular devices to offload computation tasks to cloud center, fog node or neighborhood devices by restructuring the network connection and forwarding the data flow. While some research has been carried out on green IoV and fog computing offloading [12–14], there have been few investigations into the energy-aware offloading particularly for battery-powered IoV devices. More Electric Vehicles (EVs) or hybrid EVs will be available in the transportation system for their benefits over conventional vehicles in terms of less harmful emissions. Besides, there are plenty of battery-equipped RSUs already deployed in the system. Once the vehicular device is on low battery power mode, it may automatically shut down or turn off the communication module. In this case, the remaining application will not be able to complete [15, 16]. In order to extend the battery life of the edge devices to last longer, the applications on the edge devices can be appropriately transferred to fog devices, cloud or other edge devices with enough resources and battery power.

In this chapter, SDN and fog computing are applied to the IoV to support computation offloading. A dynamic edge device offloading scheme is designed to prolong the running time of the IoV system using the remaining battery power of the edge devices to execute more applications. The remaining battery power is defined as a dynamic weight factor in the execution cost model in order to adjust the optimization objective. The execution cost model in this work identifies the energy consumption of edge devices and the overall response time of interdependent applications. Specifically, the execution cost model is defined as the weighted summation of the response time and energy consumption. With the decrease of the battery power of the edge devices, the impact of the energy consumption in the cost model will intensify. Therefore, the constraints of the overall execution response time can be appropriately relaxed to extend the battery life by offloading some applications to the fog devices or cloud. So the IoV system equips with this offloading scheme would dynamically adapt to the changing of the energy consumption. When the edge device is fully charged, the main goal is to minimize the execution delay.

As the edge device's battery power is gradually reduced, the optimization objective increasingly switches to the edge device's local energy consumption to extend the edge device's battery life and satisfy the user's execution delay requirement. A new heuristic optimization algorithm using the minimized execution cost model as the searching condition is also designed to find the appropriate target execution device for the applications on edge devices. The experimental results show that the proposed offloading approach could effectively prolong the running time of the simulated system. It should be addressed that the proposed approach is different from existing approaches in different aspects. First, the execution cost model could dynamically adjust the weight of the battery power of the device to find the balance between providing efficient service and keeping the system alive. The majority of the energy-aware approaches only focus on minimizing the energy consumption without considering the user experience. Second, our approach modeled the dependency between applications which is practical in real-world scenarios.

5.2 System Model

We employ a SDN and Fog computing empowered hierarchical system architecture of IoV, as shown in Fig. 5.1. Multiple edge devices will collaborate in the local edge network. These edge devices may connect to the fog node or connect to the cloud center directly [17]. During the execution of the application, an edge device can offload the task to other nodes with the consideration of the constraints of the application in order to improve the performance. The battery power of the equipment on the bus in Fig. 5.1 is too low to finish the three remaining tasks. The offloading scheme will evaluate the fog environment and the dependency of the application to offload the application to the bus station with the help of the SDN controller. Before discussing the formulated system model, some notations and concepts are defined and clarified first.

Definition 5.1 (Device, Local Device and Offloading Device) Devices: All electronic devices of IoVs that can perform applications, including the vehicle, the equipment in the vehicle, the SEVs, fog devices, and cloud centers.

Whenever an application is executed, it is not a primary concern what device it is executed on, but the response time and the energy consumption as a result of

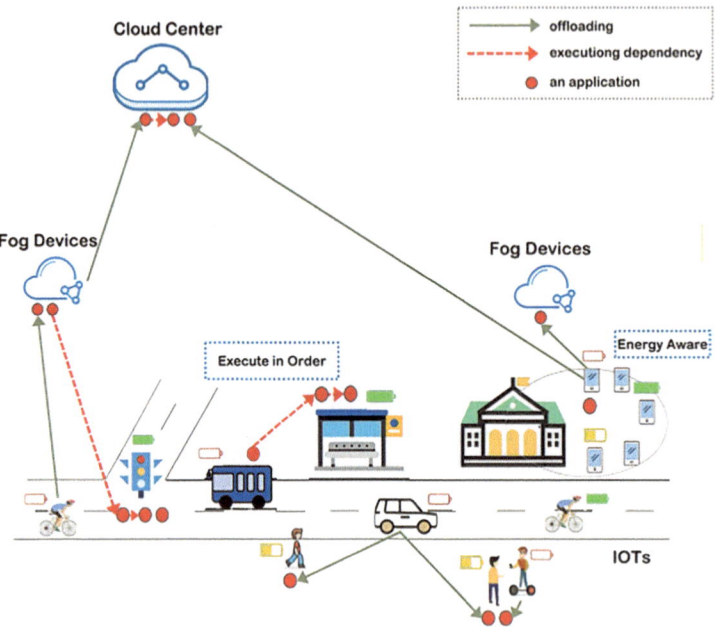

Fig. 5.1 Energy aware offloading in SDN-based IoV

this process. Therefore, this chapter does not focus extensively on the device the application is executed on but focuses more on the statistics received as a result of this execution process.

Local Devices and Offloading Devices There are many edge devices under the same WLAN, those of which are connected to multiple fog devices and cloud centers. For an application, the devices that the application requests are local devices, and the other devices that receive such applications from local devices are offloading devices. The local device can offload the local application to the offloading device, and the offloading device assists the local device in completing the execution of the application.

Definition 5.2 (Local Energy Consumption, Transmission Energy Consumption, Offloading Device Energy Consumption, and Total Execution Energy Consumption)

Local Energy Consumption refers to the energy consumed by the local device during application execution.

Transmission Energy Consumption refers to the energy consumed during the transfer of the application from the local device to the offloading device.

Offloading Device Energy Consumption refers to the energy consumed by the offloading device while executing the application.

Total Execution Energy Consumption is the sum of the local energy consumption, transmission energy consumption, and offloading device energy consumption.

Table 5.1 Description of notations

Notation	Definition
D	A group of devices connected in the same WLAN
M	The numbers of devices of the group D
d_s	The local device of the group D
d_m	The offloading device of the group D, m = 1, 2, 3, ..., M
f	The device's frequency
l	The device type(edge device or fog device)
e	The device's remaining battery capacity
p	The device's current CPU utilization
a_{ns}	The n-th application executed on the device d_S
t^d	The maximum execution tolerance delay time
θ	The amount of data to be processed
ω	The required CPU cycles to calculate the application
r $r + T^{req}$	A group of predecessor applications of the application the request time of the application
Z_{nm}	The cost of the applications a_n execution on the device d_m
q_{nm}	The weight to balance the cost Z_{nm} and the battery life of ED
E_{nm}	The energy consumption of the a_n execution on the d_m
t_{nm}	The execution slot of the a_n execution on the d_m
TE_{nm}	The completion time for the a_n execution on the d_m
TS_{nm}	The start time for the a_n execution on the d_m
t_{nm}^{tran}	The transfer slot for the a_n execution on the d_m
E_{nm}^{tran}	The transmission energy consumption for the a_n execution on the d_m
I	The offloading solution generated by the algorithm

When the application is executed entirely by the local device, both the transmission energy consumption and the offloading device energy consumption are zero.

Suppose there are M devices in a device group D connected in the same WLAN. The device is represented as $d(f, l, e, p)$. In this group D, one local device is d_s, $s \in \{1, 2, 3, ..., M\}$, then the offloading candidate devices are represented as d_m, ($m = 1, 2, 3, ..., M \& m \neq s$). The descriptions of the notations used in this chapter are given Table 5.1.

In the Table 5.1, if there are N applications running in group D, we define $a_{nm} \left(t^d, \theta, \omega, r, T^{req}\right)$, $n \in \{1, 2, 3, ..., N\}$ as the n-th application executed on the device d_m. The dependency among the applications is modeled by a directed cyclic graph as discussed in [18]. There are two execution modes for each application: local execution mode and offloading execution mode. Local execution means the application is executed on the original edge device. The offloading execution mode is to offload the application to another device (offloading device) to reduce energy consumption and at the same time meets the dependency and response time constraints.

5.2.1 Local Execution

For an application a_n executing on the device d_s locally, we define the cost of local execution is Z_{local} :

$$Z_{\text{local}} = Z_{ns} = ((1 - q_{ns}) E_{ns} + q_{ns} T E_{ns}) \tag{5.1}$$

The q_{ns} represents the impact of the local battery power consumed of the application a_n executed the device d_s on the d_s battery power.

$$q_{ns} = \frac{E_{ns}}{e_s} \tag{5.2}$$

e_s is the current remaining battery capacity of device d_s. E_{ns} is the local execution energy consumption of the application a_{ns} at device d_s.

$$E_{ns} = k\omega_n f_s (1 - p_s)^2 \tag{5.3}$$

And k is the effective switched capacitance relying on the chip architecture, k $=$ 10^{-28} as described in [19, 20].

The $T E_{ns}$ is the completion time, which equals the start time $T S_{ns}$ plus the local execution time t_{ns}:

$$T E_{ns} = T S_{ns} + t_{ns} \tag{5.4}$$

The starting time of the execution of application a_n on device d_s is defined as $T S_{ns}$, whereas the completion time is defined by $T E_{ns}$. In order to consider the execution dependencies of applications in the system model, this chapter transforms the execution dependencies into the constraint relationships of execution time similar to [21]. Figure 5.2 precisely describes the time related concepts used in our system model.

When the application is executed on the local device, its predecessor applications should have been finished in advance. All the applications in the application group r can only be executed when the predecessor applications complete their execution. Therefore, the start-time $T S_{ns}$ is:

$$T S_{ns} = \max_{i=1,2,3,\dots,M;\, j \in r} T E_{ij} \tag{5.5}$$

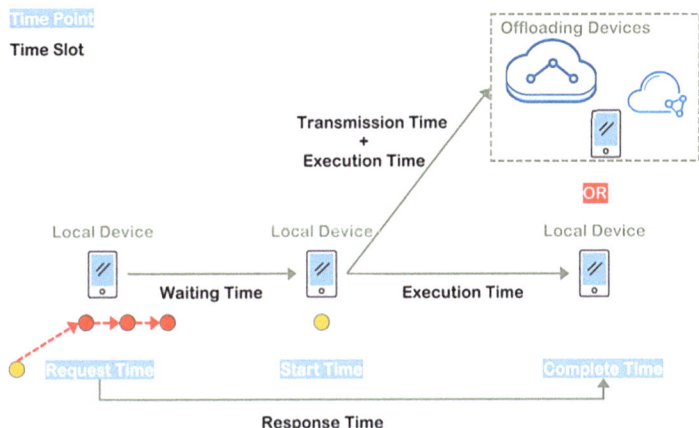

Fig. 5.2 The time model of task execution

$T E_{ij}$ is the completion time when the application a_j in the group r is executed on $d_i (i = 1, 2, 3, \ldots, M)$. And the local execution slot t_{ns} could be estimated as follows:

$$t_{ns} = \frac{\omega_n}{f_s (1 - p_s)} \tag{5.6}$$

5.2.2   Offloading Execution

When the available resource provided by the local device fails to meet the requirement of the application, the device will try to transfer the application to another device d_m for execution. Offloading execution includes two main steps: data transmission and execution. Let us assume t_{nm}^{tran} to represent the transfer delay of the application a_n to the offloading device d_m:

$$t_{nm}^{\text{tran}} = \frac{\theta_{nm}}{R_{nm}} \tag{5.7}$$

where R_{nm} represents the data transfer rate at this time, which is calculated using the Shannon's Theorem [22]:

$$R_{nm} = B_m \log_2 \left(1 + \frac{P_{nm} H_{nm}}{\sum_{i \neq m, bj \neq n} P_{ij} H_{ij} + \sigma^2} \right) \tag{5.8}$$

B_m indicates the bandwidth of the transmission link. When the type of device d_m is $l = 0$, $B_m = B_{\text{wlan}}$; when the type of device d_m is $l = 1$, $B_m = B_{\text{wan}}$. The B_{wlan} is the bandwidth of WLan and the B_{wan} is the bandwidth of Wan. P_{nm} is transferred power when the application a_{ns} is transferred to the device d_m, H_{nm} is the gain, and σ^2 is the noise ratio.

Use E_{nm}^{tran} to indicate the transmission energy consumed by the application a_{ns} to the offloading device d_m:

$$E_{nm}^{\text{tran}} = P_{nm} t_{nm}^{\text{tran}} \qquad (5.9)$$

For the offloading execution, we define its cost is Z_{off}:

$$Z_{off} = Z_{nm} = ((1 - q_{nm}) \, E_{nm} + q_{nm} T E_{nm}) \qquad (5.10)$$

The local battery power consumed of offloading execution is E_{nm}^{tran}, So the q_{nm} is $\frac{E_{nm}^{tran}}{e_s}$. The total execution energy E_{nm} of the application a_n on the offloading device d_m is:

$$E_{nm} = k\omega_n f_m \, (1 - p_m)^2 + E_{nm}^{\text{tran}} \qquad (5.11)$$

It is the energy consumption of application a_n executed on the offloading device d_m plus the transmission energy consumed. The total execution time t_{nm} of the application a_n on the offloading device d_m is:

$$t_{nm} = \frac{\omega_n}{f_m \, (1 - p_m)} + t_{nm}^{\text{tran}} \qquad (5.12)$$

It is the execution time of the application a_n on the offloading device d_m plus the transfer delay. In the offload execution mode, the application a_n can only be moved after the execution of the local execution dependent application has been completed, so the time to reach the offloading device is the time at which the application starts moving plus the transmission time. In addition, the execution of the offloading device d_m depends on the execution of the application after the execution of the application a_n, so the application a_n at the start time of the offloading device d_m 's $T S_{nm}$ is:

$$T S_{nm} = \max_{j \in r} \left\{ \max \left\{ T E_{sj} + t_{sj}^{tran} \right\}, \max_{m \neq s} \left\{ T E_{mj} \right\} \right\} \qquad (5.13)$$

$\max \left\{ T E_{sj} + t_{sj}^{tran} \right\}$ is the time for the application a_n to reach the offloading device d_m, $\max_{m \neq s} \left\{ T E_{mj} \right\}$ is the time at which the application a_n can be executed on the offloading device d_m. End time $T E_{nm}$ is the opening time plus the execution time:

$$T E_{nm} = T S_{nm} + \frac{\omega_n}{f_m \, (1 - p_m)} \qquad (5.14)$$

In this section, we formalized the system model as a group of equations. To our best knowledge, it is the first precise model to formalize the energy cost, execution time constraints and application dependencies all together. A precise theoretical model is essential to optimize to offloading of the applications.

5.3 Offloading Optimization

Based on the system model we defined in the previous section, we first formalize the offloading optimization objective functions and then propose the optimization algorithm to generate the offloading scheme. We investigate the problem into two typical scenarios: (1) the application offloading without considering other devices' battery power; (2) the offloading considers the battery power of all devices in the device group.

5.3.1 Offloading Considering Single Device Battery Power

Here we use matrix I to represent the offloading scheme, and $I_{nm} \in I$ to indicate the execution location of the application a_n:

$$I_{nm} = \begin{cases} 1, & \text{application } a_n \text{ executes on device } d_m \\ 0, & \text{otherwise.} \end{cases} \tag{5.15}$$

I_{nm} is a binary variable. When $I_{nm} = 1$, it means that the application a_n is executed on the device d_m. When $I_{nm} = 0$, the application a_n in device d_m is not executed. When the id of the requesting device is equal to the id of the local device, it means that it is executed locally, otherwise it is offloading. Thus the cost of executing a_n for each device d_m, $m = 1, 2, 3, .., M$ is Z_{nm}:

$$Z_{nm} = I_{nm} \left((1 - q_{nm}) E_{nm} + q_{nm} T E_{nm} \right) \tag{5.16}$$

Z_{nm} can dynamically adjust the execution energy consumption and response time to the execution cost according to the remaining battery power of the local device. As the amount of battery power requested is continuously reduced, the impact of execution energy consumption on execution costs becomes greater. Execution cost minimization is more common on devices that save local power while meeting user requirements for response time.

In order to maximize the user's benefit, each application will be selected to minimize the cost of execution. The final objective function is defined as follows:

$$\text{Obj-1:} \quad \min \sum_{n=1}^{N} \sum_{m=1}^{M} Z_{nm}$$

$$= \min \sum_{n=1}^{N} \sum_{m=1}^{M} I_{nm} \left((1 - q_{nm}) E_{nm} + q_{nm} T E_{nm} \right)$$

$$\text{s.t.} \begin{cases} C1: & I_{nm} \in [0, 1] \quad m = (1, 2, 3, \ldots, M), \\ C2: & \sum_{m=1}^{M} I_{nm} \leq 1, \\ C4: & T E_{nm} \leq t_n^d + T_n^{req}, \\ C5: & T S_{nm} \geq \max_n, \\ & T S_{nm} - t_{nm}^{tran} \geq \max_{j \in r} \{ T E_{ij} \}, \end{cases} \tag{5.17}$$

As the local power is reduced, the proportion of energy consumption in the objective formula becomes larger, and the impacts will become greater and greater. Intuitively, high-energy applications will be more frequently moved to offloading devices for execution, for reducing local battery energy consumption and saving the power of the local device.

The C1 to C5 are the five constraints of the objective function. C1 indicates that the application a_{ns} has only two states of execution and no execution on the device d_m, so I_{nm} is binary, equals to 1 or 0. C2 reflects that an application can only be executed on one device, or otherwise failed. C3 suggests that the execution of the application needs to meet the quality of service (QoS) of the user. The total response time of the application needs to be less than the maximum delay that the user endures. C4 determines that the remaining battery power (current power) of the local device can complete the application. The C5 constraint is derived from the formula in Eqs. (5.4), (5.5), (5.13), and (5.14), which acknowledges that the application must be executed locally before the previous application is executed. The offloading execution of the local device have to first wait for the previous application to be completed before offloading to the offloading device.

This algorithm has a time complexity of $O(MN)$, this indicates the advantage of this algorithm is that it can find the optimal execution device for each application comparatively quick for the low volume of applications or devices. However, for large-scale equipment and applications, the execution time may be too long due to excessive calculation time.

In edge computing, if the application a_n leave request device d_s to any other device, then a_n execution cost on any other device Z_{nm} must be less than Z_{ns}. The formula (5.18) is based on this conclusion, by setting up the Z_{nm} threshold of assistance equipment, screening out the assistance equipment that does not meet the threshold, so that the number of M can be reduced in our algorithm. Because each parameter in Z_{ns} is determined, Z_{ns} can be regarded as a constant C. According to the decomposition of Z_{nm}, it is found that the value of $f_m (1 - p_m)$ can be solved. In this chapter, the calculated value is marked as ΔG, which is used as the pre-processing condition at the beginning of the algorithm. The f_m of the device d_m is greater than the threshold ΔG, indicating that the application a_n can be offloaded to d_m, and vice versa.

$$Z_{ns} \geq Z_{nm}$$

$$\Rightarrow (1 - q_{ns}) E_{ns} + q_{ns} T E_{ns} \geq (1 - q_{nm}) E_{nm} + q_{nm} T E_{nm}$$

$$\Rightarrow C \geq \left(1 - \frac{E_{nm}^{\text{tran}}}{e_n} \right) E_{nm} + \frac{E_{nm}^{\text{tran}}}{e_n} T E nm$$

$$\Rightarrow C \geq \left(1 - \frac{P_{nm} \frac{\theta_{nm}}{R_{nm}}}{e_S} \right) k\omega_n f_m (1 - p_m)^2 + P_{nm} \frac{\theta_{nm}}{R_{nm}} e_s$$

$$+ \frac{P_{nm} \frac{\theta_{nm}}{R_{nm}}}{e_s} \left(\max_{j \in r} \left\{ \max \left\{ T E_{sj} + t_{sj}^{tran} \right\}, \max_{m \neq s} \left\{ T E_{mj} \right\} \right\} \right.$$

$$\left. + \frac{\omega_n}{f_m (1 - p_m)} \right)$$

$$\Rightarrow f_m (1 - p_m) \geq \Delta G \qquad\qquad (5.18)$$

5.3.2   Offloading Considering Multiple Devices' Battery Power

The multi edge node energy aware load transfer decision-making algorithm is based on the single edge node scenario. It takes into account the resource contention of the applications on each edge device for the global available power. The goal is to extend the power of all edge devices in the system to minimize the execution response time and energy consumption as the constraints to find the best assistance device group. The decision algorithm uses the idea of heuristic search, taking Z_{nm} as the actual cost in the evaluation function and $\frac{E_{nm}^{tran}}{e_{ns}} + \frac{E_{nm}}{e_{nm}}$ as the estimated cost of the evaluation function. The specific meaning is that when application a_n is executed on device d_m, the consumed energy accounts for the power consumption of the equipment involved. If $m = s$, then $E_{nm}^{tran} = 0$; if $m \neq s$, then $\frac{E_{nm}^{tran}}{e_{ns}}$ represents the proportion of transmission energy consumption to the remaining power of local equipment, and $\frac{E_{nm}}{e_{nm}}$ represents the proportion of execution energy consumption on device d_m to the remaining power of device d_m. On the premise that the global device power does not increase and is only consumed by the application if every time $\frac{E_{nm}^{tran}}{e_{ns}} + \frac{E_{nm}}{e_{nm}}$ is the minimum, then the application execution energy consumption accounts for the smallest proportion of the overall remaining power, then the selected scheme must make the global device last for the longest time. Therefore, in the energy-aware load transfer decision algorithm of multi-edge nodes, the target formula for a certain application a_n is shown as follows:

$$f(x) = g(x) + h(x) \Rightarrow F_{nm} = Z_{nm} + \frac{E_{nm}^{tran}}{e_{ns}} + \frac{E_{nm}}{e_{nm}} \tag{5.19}$$

$$Obj - 2: \quad \min \sum_{n=1}^{N} \sum_{m=1}^{M+1} F_{nm} = ff \min \sum_{n=1}^{N} \sum_{m=1}^{M+1} I_{nm} \left(Z_{nm} + \frac{E_{nm}^{tran}}{e_{ns}} + \frac{E_{nm}}{e_{nm}} \right)$$

$$\text{s.t.} \quad \begin{cases} C1 - C5, \\ C6: \quad e_{nm} \geq E_{nm}. \end{cases} \tag{5.20}$$

The constraint $C1 - C5$ are from the objective function $Obj - 1$. In addition, a new constraint $C6$ needs to be added. When the application a_n on the device is requested to execute on the offloading device d_m, the remaining power e_{nm} of the assisting device can support the energy consumption required for execution. In the situation of multi edge nodes, because the target formula has changed, so the calculation of ΔG also needs to be updated. The calculation formula is changed to:

$$Z_{ns} + \frac{E_{ns}}{e_{ns}} \geq Z_{nm} + \frac{E_{nm}^{tran}}{e_{ns}} + \frac{E_{nm}}{e_{nm}} \tag{5.21}$$

In heuristic search, the proportion of energy consumption in the global equipment power consumption of a_n is calculated on device d_m and is used as the estimated cost function. If the proportion is the smallest, it is proved that the global power consumption of application a_n is the smallest when it is executed on device d_m, and the more power resources left, the more resources available for the remaining unexecuted applications. Therefore, it is very possible to find the optimal solution by using the evaluation function mentioned in this chapter. The pseudo-code is shown in Algorithm 1. The heuristic search algorithm first calculates the computing capacity threshold ΔG according to Eq. (5.21) and uses this parameter to filter the global equipment to screen out the equipment with insufficient computing power. Then, based on Eqs. (5.3), (5.4), (5.11), and (5.14), the execution energy consumption, the execution response time of application a_n and the offloading device d_m are calculated respectively. According to the objective function (5.17), the value of Z_{nm} is calculated, and the device d_m which makes Z_{nm} the smallest is the result of the calculation. In addition, we also need to calculate the execution equipment of other programs, and the combination of the execution devices of each application is the decision result.

Algorithm 5 HBLB algorithm

Require: s, M, N
Ensure: $I_{N \times M}$
 1: **function** HEURISTIC SEARCH(s, M, N)
 2:     $I_{N \times M} = 0$
 3:     **for** $n = 1 \to N$ **do**
 4:         $F_{\min} = \infty$
 5:         Computation ΔG
 6:         **for** $m = 1 \to M + 1$ **do**
 7:             **if** $f_m(1 - p_m) \geq \Delta G$ **then**
 8:                 **if** $m = s$ **then**
 9:                     Computation t_{nm}, E_{nm}, TS_{nm}
10:                 **else**
11:                     Computation $t_{nm}^{tran}, E_{nm}^{tran}$
12:                     Computation t_{nm}, E_{nm}, TS_{nm}
13:                 **end if**
14:                 Computation TE_{nm}
15:                 Computation F_{nm}
16:                 **if** $F_{nm} \leq F_{\min}$ **and** $TE_{nm} \leq t_d^n + T_{req}^n$ **then**
17:                     $Z_{\min} = Z_{nm}$
18:                     $o = m$
19:                 **end if**
20:             **end if**
21:         **end for**
22:         $I_{no} = 1$
23:         **if** $o == s$ **then**
24:             $e_{(n+1)s} = e_{ns} - E_{ns}$
25:         **else**
26:             $e_{(n+1)s} = e_{ns} - E_{no}^{tran}$
27:             $e_{no} = e_{no} - E_{no}$
28:         **end if**
29:     **end for**
30:     **return** $I_{N \times M}$
31: **end function**

5.4 Implementation

5.4.1 Simulation Setup

To evaluate the performance of the algorithm, we use two types of equipment as experimental devices: edge devices with weak processing power as well as limited memory storage; and fog devices with stronger processing power and moderate memory storage.

We prepared 4 Orange Pi Pc Plus and 4 Raspberry Pi 3 Model B as edge devices (like the RSUs), and 2 ASUS vm590L as fog devices (like the base station). The specific parameters are shown in Table 5.2. The communication bandwidth between two connected edge devices are $B_s = 100\,\text{Mb/s}$, the communication bandwidth between an edge device and a fog device is $B_f = 50\,\text{Mb/s}$, and the gain in the

Table 5.2 Devices in the experiment

Device type	Unit type	Hardware specification	Device ID	CPU utilization
Edge device	Orange Pi Pc Plus	H3 Quad-core Cortex-A7 H.265/HEVC 4K CPU	1	0.3
		Mali400MP2 GPU @600MHz	2	0.5
		Supports OpenGL ES 2.0	3	0.6
		Memory (SDRAM) 1GB DDR3 (shared with GPU)	4	0.3
	Raspberry Pi 3 Model B	Quad Core 1.2GHz Broadcom BCM2837 64bit CPU	5	0.6
		1GB RAM	6	0.1
		BCM43438 wireless LAN	7	0.1
		100 Base Ethernet	8	0.4
Fog device	ASUS VM590L	Intel Core i7-4510U 2.4GHz CPU	9	0.1
		4GB DDR3L 1600MHz	10	0.6

channel is $H_{sm} = K_{s,m}^{\xi}$, where $K_{s,m}$ is the distance between the device d_s and the device d_m, and ξ is the path loss factor. The power P of the channel is estimated based on the wireless network specification and set to 1 W as many research has been done [23]. Because the cost Z_{mn} is the sum of time and energy, the dimension is not unified. In order to solve this problem, the zero menu normalization is used with the experimental samples prepared in the experimental environment.

The device d_1 is the local device, that is, the requesting device, which has a set of programs $a_n (n = 1, 2, \ldots, N)$. The rest of the devices of group D are offloading devices that help the local device complete the application group $a_n (n = 1, 2, 3, \ldots, N)$. In order to emulate the performance of our algorithm, three comparison algorithms are proposed:

The Local Algorithm The device d_1 executed all applications of its group $a_n (n = 1, 2, 3, \ldots, N)$ locally using its own CPU. When the device d_1 's battery is exhausted, the application's execution stops.

The Offloading Algorithm All applications of the group $a_n (n = 1, 2, 3, \ldots, N)$ on d_1 are executed on offloading devices $d_m (m \neq 1)$. The objective of every application executed on the offloading device is:

$$\min \sum_{n=1}^{N} \sum_{m=2}^{M} I_{mn} (E_{mn} + T E_{mn}) \qquad (5.22)$$

The constraints C1–C5 above mentioned need to be considered. The objective function is used to find the offloading device that saves the most response time and total execution energy.

The Fixed-Weight Algorithm This algorithm is frequently adopted by many application offloading algorithms [24]. It represents the algorithms use fixed weight ratio for execution time and energy consumption. The fixed-weight algorithm is similar to the algorithm used in this chapter, where both applications can be executed on local or the offloading devices. However, the objective function of the fixed-weight algorithm does not take into account the dynamics of power of the local device: as the application is executed, the power of the local device is gradually decreasing. The objective is as follows:

$$\min \sum_{n=1}^{N} \sum_{m=1}^{M} I_{mn} \left((1 - w) E_{mn} + w T E_{mn} \right) \tag{5.23}$$

The fixed-weight algorithm uses a fixed weight to adjust the overall execution time and the total execution energy consumption.

5.4.2 Performance Analysis

In this section, we conducted series of experiment to validate the efficiency of our proposed solution. This approach is aimed to handle failures, task execution interruptions, delays etc. by implementing energy-aware offloading for energy-constraint edge devices. To facilitate a comprehensive analysis, we have carried out four sets of comparative experiments that demonstrate energy-aware application offloading algorithm is effective in different testing conditions. These experiments are outlined in the following sections. The experiment results marked as "Single edge device" refers to the algorithm proposed in Sect. 5.3.1, which only considers the battery power of the local device, whereas the "HBLB" refers to the algorithm that considers all devices' battery power.

System Resilient Under a Given Battery Constraint
In this experiment, we validated the resilient of our proposed offloading scheme by evaluating the number of applications that can be successfully executed as the battery power gradually decline. As shown in Fig. 5.3a, the x-axis represents the percentage of the initial battery power of the local device, and the y-axis represents the number of the application completed when the battery of the local device is exhausted. Each point represents an independent and complete experiment: from the initial power represented by x-axis to 0, the number of applications completed by the edge devices is showed on y-axis. Figure 5.3b and c did the same experiments besides the total number of applications is changed to 150 and 500 respectively. Compared with the local and the offloading algorithm, the proposed HBLB offloading algorithm could finish more applications before the battery is exhausted with the different percentages of the initial battery power.

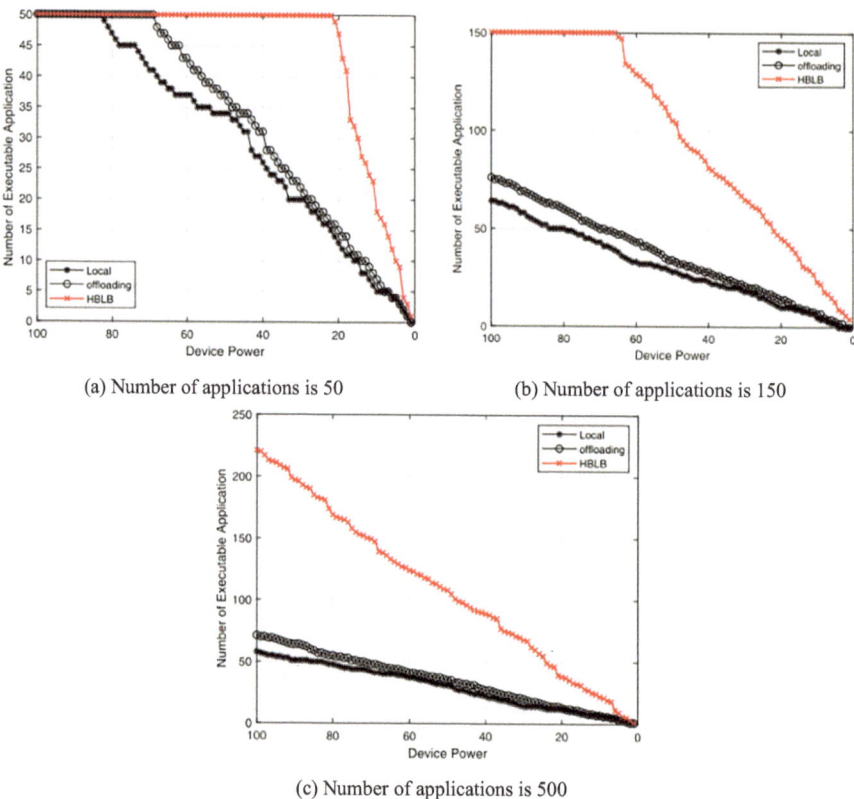

(a) Number of applications is 50   (b) Number of applications is 150

(c) Number of applications is 500

Fig. 5.3 The number of applications could execute with different initial battery power. (**a**) Number of applications is 50. (**b**) Number of applications is 150. (**c**) Number of applications is 500

More importantly, it execute almost the same number of applications as the battery percentage gradually decline from 100 to about 20%. A decrease in battery power to a certain percentage has minimal impact of the system performance. Thus, the system becomes less sensitive to the power consumption which would have otherwise, adversely affect and in some case interrupt the normal; operation of the device. This further validates the advantages of incorporating stability and resilient in edge computing environment. A similar performance is achieved despite increasing the when the number of application to 150 as shown in as in Fig. 5.3b. The applications in each test group are randomly generated with different data sizes, computations, and dependencies. This proofs that HBLB algorithm could generate optimized offloading schemes under different execution environments. Furthermore, it validates the significance of building a resilience for edge computing to enhance the performance of devices and application which heavily rely on energy.

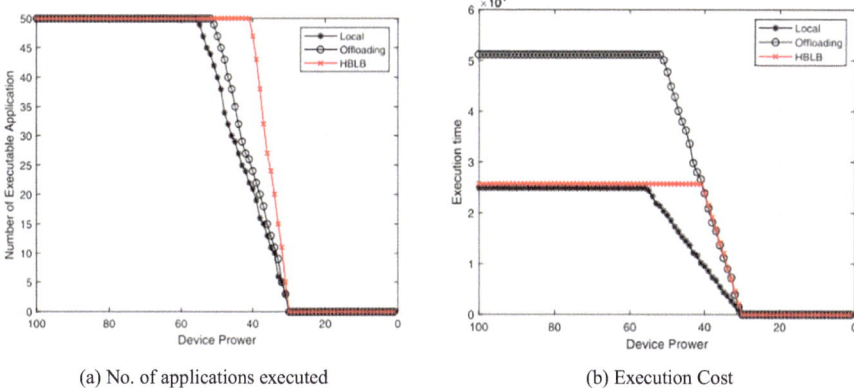

(a) No. of applications executed

(b) Execution Cost

Fig. 5.4 Evaluating the execution cost for the case of sufficient device power. (**a**) No. of applications executed. (**b**) Execution cost

5.4.3 Cost of Energy-Aware Offloading

In this experiment, a new group of 50 applications was generated randomly. Figure 5.4a shows the number of applications that the local device could execute exhausting the battery power. As shown in Fig. 5.4a, when the initial battery power is sufficient (greater than 60%), a device using any of the algorithms is capable of executing and finishing all the 50 applications. In order to critically analyse the cost of the HBLB system, it is imperative to first assess the local device's performance without offloading. The local device is evaluated for a case of sufficient battery power, hence the experiment is stopped when the batter power is lower than 30%.

Figure 5.4b shows the total execution cost incurred by the offloading which is a function of total execution time and total energy consumption of the entire system. The results shows that our algorithm introduced some overhead to the system because our algorithm needs to calculate the optimized offloading scheme. But in real-life, the battery power of the device is usually not enough to execute all applications. Since the main objective of this offloading strategy is to slightly increase energy consumption and computation to find the optimized offloading scheme to offload the applications to a more suitable device to finish more applications in the device group. In effect, the overall power consumption of the system increases to sustain the task execution. Consequently, it has negative effect on the overall cost. This is an acceptable trade-off to achieve the objective continuous task execution in the event of low battery power for latency-aware and energy-dependent applications. Additionally, all the execution time and energy related constraints are still satisfied to guarantee the offloading scheme meets the users' requirements.

We also conducted a comparative experiment to evaluate the effectiveness of the proposed energy-aware offloading approach when the initial power is insufficient. As shown in Fig. 5.5a, the device's initial battery power was set to half of what it was in the previous experiment (Fig. 5.4a). While the overall cost of our solution

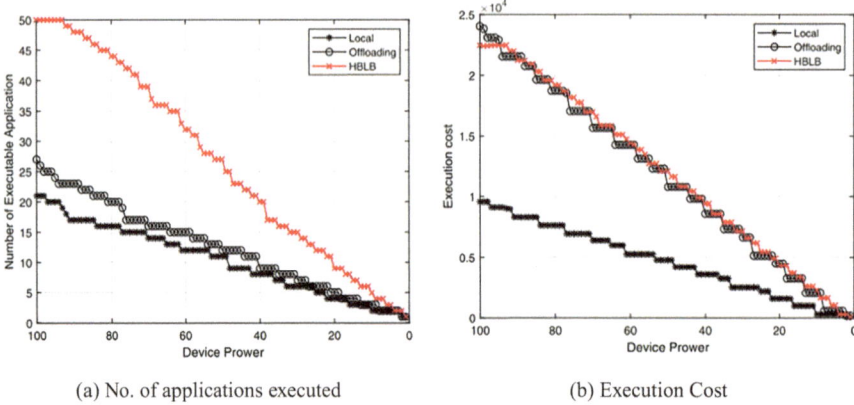

(a) No. of applications executed   (b) Execution Cost

Fig. 5.5 Comparing the performance of HBLB algorithm with the local algorithm and the offloading algorithm on the local device without enough battery power. (**a**) No. of applications executed. (**b**) Execution cost

remains slightly higher than other algorithms, it demonstrates the resilience needed for real-time applications. Specifically, although our solution maintains a similar execution cost compared to other offloading algorithms, it significantly outperforms them in the number of applications executables on low-power devices. This is because our algorithm optimally leverages the battery power of devices within the fog network, offloading tasks to the most suitable device. This scenario illustrates practical applications where many devices have lower power capabilities.

5.4.4 Comparison with Fixed-Weight Algorithm

This section presents a comparison of the proposed algorithm with the fixed-weight algorithm as shown in Fig. 5.6. The local device is set with the same battery power and assigned the same application group includes 150 applications. The experiment results shows that the Single edge device algorithm could execute slightly more applications than the fixed-weight algorithm while the two algorithms performs poorly compared to the HBLB algorithm. The poor performance of the fixed-weight algorithm is attributed to the fact that: it does not consider the dynamic power consumption of the device. Instead, it uses fixed weight to adjust the execution time and energy consumption. As a result, it performs poorly in realistic applications. Hence, it cannot guarantee system resilience in certain applications. Contrary to this approach, the HBLB algorithm consider the dynamic power of local devices as well as the battery power of all the devices in the network for efficient selection of the offloading device. Consequently, the number of task and applications and successfully executed to meet the latency requirement. Most importantly ensure uninterrupted and continuous task execution which further justifies the role of energy-aware offloading in edge computing resilience.

Fig. 5.6 Comparison with fixed-weight algorithm

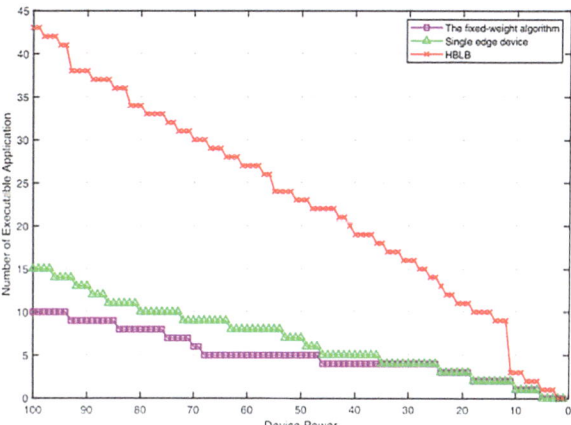

5.5 Conclusion

In this chapter, we studied and analyzed the design of resilient system in edge computing by implementing an energy-aware offloading for constraint devices. The aim of the proposed scheme is build a resilient system which ensure continuous uninterrupted task execution for battery dependent devices. A comprehensive system model is proposed to formalize the time constraints, energy consumption, and application dependencies. This system model and the cost model established on the system model make our work different from the majority of the existing works. When designing the algorithms to find the optimal offloading scheme, two typical application scenarios are discussed respectively to make the algorithm more adapt to real-life applications.

Experiments show that the offloading algorithm proposed in this chapter could migrate some applications to more proper devices to finish more applications from the system's point of view. Our solution could increase the number of executed applications significantly compared with the system without offloading. It also performs better than some typical solutions like fixed-weight algorithms. Some more advanced algorithms such as machine learning algorithms may help us find the optimal offloading scheme without bringing in high overhead.

References

1. Liu, K., Xu, X., Chen, M., Liu, B., Wu, L., Lee, V.C.S.: A hierarchical architecture for the future Internet of vehicles. IEEE Commun. Mag. **57**(7), 41–47 (2019)
2. Hou, X., et al.: Reliable computation offloading for edge-computing-enabled software-defined IoV. IEEE Internet Things J. **7**(8), 7097–7111 (2020)
3. Kadhim, A.J., Hosseini Seno, S.A.: Maximizing the utilization of fog computing in Internet of vehicle using SDN. IEEE Commun. Lett. **23**(1), 140–143 (2019)

4. Huang, C., Chiang, M., Dao, D., Su, W., Xu, S., Zhou, H.: V2v data offloading for cellular network based on the software defined network (SDN) inside mobile edge computing (MEC) architecture. IEEE Access **6**, 17741–17755 (2018)
5. Suyama, T., Kishino, Y., Naya, F.: Abstracting IoT devices using virtual machine for wireless sensor nodes. In: Proc. IEEE World Forum Internet Things (WF-IoT), Mar. 2014, pp. 367–368 (2014)
6. Heo, S., Woo, S., Im, J., Kim, D.: IoT-MAP: IoT mashup application platform for the flexible IoT ecosystem. In: Proc. 5th Int. Conf. Internet Things (IOT), Oct. 2015, pp. 163–170 (2015)
7. Du, X., Shayman, M., Rozenblit, M.: Implementation and performance analysis of SNMP on a TLS/TCP base. In: Proc. IEEE/IFIP Int. Symp. Integr. Netw. Manage., pp. 453–466 (2001)
8. Guo, H., Zhang, J., Liu, J., Zhang, H., Sun, W.: Energy-efficient task offloading and transmit power allocation for ultra-dense edge computing. In: Proc. IEEE Global Commun. Conf., Dec. 2018, pp. 1–6 (2018)
9. Guo, H., Liu, J., Zhang, J.: Computation offloading for multi-access mobile edge computing in ultra-dense networks. IEEE Commun. Mag. **56**(8), 14–19 (2018)
10. Varshney, P., Simmhan, Y.: Demystifying fog computing: Characterizing architectures, applications and abstractions. In: Proc. IEEE 1st Int. Conf. Fog Edge Comput. (ICFEC), May 2017, pp. 115–124 (2017)
11. Shi, Y., Chen, S., Xu, X.: MAGA: A mobility-aware computation offloading decision for distributed mobile cloud computing. IEEE Internet Things J. **5**(1), 164–174 (2018)
12. Ning, Z., et al.: Deep reinforcement learning for intelligent Internet of vehicles: An energy-efficient computational offloading scheme. IEEE Trans. Cogn. Commun. Netw. **5**(4), 1060–1072 (2019). https://doi.org/10.1109/TCCN.2019.2930521
13. Chen, H., Zhao, T., Li, C., Guo, Y.: Green Internet of vehicles: Architecture, enabling technologies, and applications. IEEE Access **7**, 179185–179198 (2019). https://doi.org/10.1109/ACCESS.2019.2958175
14. Baum, M., Dibbelt, J., Pajor, T., Sauer, J., Wagner, D., Zündorf, T.: Energy-optimal routes for battery electric vehicles. Algorithmica **82**(5), 1490–1546 (2020). https://doi.org/10.1007/s00453--01900655-9
15. Molzahn, D.K., et al.: A survey of distributed optimization and control algorithms for electric power systems. IEEE Trans. Smart Grid **8**(6), 2941–2962 (2017)
16. Bedi, G., Venayagamoorthy, G.K., Singh, R., Brooks, R.R., Wang, K.-C.: Review of Internet of Things (IoT) in electric power and energy systems. IEEE Internet Things J. **5**(2), 847–870 (2018)
17. Lin, Y.-H., Huang, J.-J., Fan, C.-I., Chen, W.-T.: Local authentication and access control scheme in M2M communications with computation offloading. IEEE Internet Things J. **5**(4), 3209–3219 (2018)
18. Dey, A., Stuart, K., Tolentino, M.E.: Characterizing the impact of topology on IoT stream processing. In: Proc. IEEE 4th World Forum Internet Things (WF-IoT), Feb. 2018, pp. 505–510 (2018)
19. Kakerow, R.: Low power design methodologies for mobile communication. In: Proc. Int. Conf. Comput. Des., Sep. 2002, pp. 8–13 (2002)
20. Burd, T.D., Brodersen, R.W.: Processor design for portable systems. J. VLSI Signal Process. Syst. **13**(2–3), 203–221 (1996). https://doi.org/10.1007/BF01130406
21. Chen, L., Wu, J., Long, X., Zhang, Z.: ENGINE: Cost effective offloading in mobile edge computing with fog-cloud cooperation (2017), arXiv:1711.01683 [Online]. Available: http://arxiv.org/abs/1711.01683
22. Verdu, S.: Fifty years of Shannon theory. IEEE Trans. Inf. Theory **44**(6), 2057–2078 (1998)
23. Mao, Y., Zhang, J., Letaief, K.B.: Dynamic computation offloading for mobile-edge computing with energy harvesting devices. IEEE J. Sel. Areas Commun. **34**(12), 3590–3605 (2016)
24. Bozorgchenani, A., Tarchi, D., Corazza, G.E.: An energy-aware offloading clustering approach (EAOCA) in fog computing. In: Proc. Int. Symp. Wireless Commun. Syst. (ISWCS), Aug. 2017, pp. 390–395 (2017)

Chapter 6
Optimization for Edge Computing

Abstract This book chapter addresses the dynamic environment of edge computing's imperative optimization area. Understanding the aims and methods of optimization is essential given the spread of edge devices and the complexity nature of applications. Task offloading, edge caching, network optimization, resource allocation, and privacy-aware optimization are just a few of the optimization strategies covered in this chapter. In-depth information on the particular difficulties and potential solutions for using these methods in edge computing settings is provided in each subsection. The chapter also examines the complex issues of edge optimization in situations with heterogeneous devices and the idea of edge node clustering for improving system performance. The chapter also discusses assessment and performance analysis tools to guarantee the viability and efficacy of certain optimization measures. It goes through how important tools for profiling, benchmarking, tracing, containerization, and the newly developed field of edge analytics are for determining how efficient and effective edge computing systems are. By the end of this chapter, readers will have a thorough grasp of the complex field of edge computing optimization, enabling them to create and put into use edge systems that are very effective and quick.

Keywords Task offloading · Edge caching · Network optimization · Resource allocation · Privacy-aware optimization · Edge node clustering · Performance analysis tools

6.1 The Goal of Optimization in Edge Computing Environment

The key goal of edge computing optimization is to fill the breach between the growing demand for high-quality, low-latency services and the physical constraints of resource-constrained devices, bandwidth-constrained networks, and power-inefficient processes. In essence, it aims to reduce latency, power consumption, and cost while making the most use of the already available compute, network, and storage options. For example, in the case of a cloud gaming platform that uses

© The Author(s), under exclusive license to Springer Nature Singapore Pte Ltd. 2024   95
Y. Zhai et al., *Edge Computing Resilience*, SpringerBriefs in Computer Science,
https://doi.org/10.1007/978-981-97-6998-8_6

edge computing, for instance, the game data is processed near the user at the edge. Here, minimizing latency, which adversely affects the gaming experience, is the main objective. The gaming platform must run smoothly even in situations where the network is congested by simultaneously using optimization strategies to limit bandwidth utilization, lower costs, and ensure that the platform is not overloaded.

6.2 Optimization Techniques

6.2.1 Task Offloading

An essential optimization method in edge computing is task offloading, commonly referred to as computation offloading. It entails moving computationally demanding tasks from resource-limited devices (such as IoT or mobile devices) to more potent systems located at the network edge. By enabling faster processing close to the data source, this technology not only helps conserve limited resources on the original devices but also lowers latency.

When implementing task offloading in an edge computing environment, there are a number of factors and concerns to take into account, such as decision-making, execution models, offloading objectives, and obstacles.

- **Offloading decision making:** The choice to offload a task is impacted by several factors, which include the task's type, the originating device's processing power, the network environment, and the edge server's availability and capabilities. Offloading can improve speed and conserve device resources, but there are trade-offs: it may also raise communication costs and expose data to security threats. To make these offloading decisions, algorithms, and models like Markov decision processes and game theory are frequently used. For instance, a smart city application may have significant processing needs for an image identification task from a security camera system. The system may opt to transfer this work to an edge server with a GPU after weighing these considerations, allowing for quicker processing and real-time response.
- **Offloading execution models:** Task offloading primarily uses two types of execution models: complete offloading and partial offloading. Partial offloading divides the task into multiple subtasks, some of which are processed on the original device and others on the edge server. Full offloading entails shifting the fully assigned job to an edge server. The details of the work at hand and the design of the system will determine whether to offload completely or partially. For instance, in a healthcare monitoring system, pre-processing of sensor data might be locally done (partial offloading) before the complicated analysis is fully offloaded to the edge server, balancing resource utilization and performance. For instance, in a healthcare monitoring system, pre-processing of sensor data might be done locally (partial offloading) before the complicated analysis is fully offloaded onto the edge server, balancing resource utilization and performance.

- **Offloading targets:** Observing the task essentials in accordance with the edge servers' capabilities, tasks may be offloaded to several types of edge servers. From robust edge data centers to less competent edge devices like routers, gateways, or even other end-user devices, this can be the case. Basic functions like data filtering in a connected car scenario might be delegated to an in-vehicle edge device, while more computationally expensive jobs like real-time traffic prediction might be delegated to a nearby edge data center with higher processing capabilities.
- **Challenges:** Despite its advantages, task offloading has a number of drawbacks. The performance of offloading might be impacted by network instability because tasks and data must be transferred between devices. Security and privacy are also major issues, particularly with regard to sensitive data. Offloading benefits may be limited by resource conflicts on edge servers, necessitating efficient resource management techniques.

6.2.2   Edge Caching

Edge data caching is a key optimization approach in edge computing that stores commonly accessed data close to the end users in order to improve data accessibility, decrease network congestion, and improve user experience. Edge caching in the context of edge computing is keeping copies of the data on edge servers that are situated close to the users. Information that has been stored at an edge server is immediately given to a user upon request, eliminating the need to obtain the data from the original source, which could be a distant data center or cloud server. Edge caching includes aspects like cache management, caching techniques, and challenges.

- **Cache management:** The principal objective of edge caching seeks to manage the amount and duration of data storage (caching). It's critical to choose which data should be stored because edge servers have a finite amount of storage space. The types of data that are often suitable candidates for caching include those that are frequently accessed, pertinent to nearby users, or expensive to get from the source. A streaming provider might, for instance, cache a well-liked movie on an edge server situated in an area where there is a considerable demand for that particular movie. The data is immediately supplied from the edge cache when viewers in that area request the movie, enhancing streaming quality and lowering data retrieval costs.
- **Caching strategies:** Depending on the requirements and features of the application, various caching techniques might be used. Least Recently Used (LRU), which replaces the recently used data when the cache is full, and Least Frequently Used (LFU), which replaces the least frequently used data, are two examples of prevalent techniques. More sophisticated solutions might take into account things like data size, retrieval costs, user preferences, or predictive analytics. As an

illustration, in an IoT application, an edge server might cache sensor data using a time-based technique, saving just the most current data and deleting older data, delivering the most recent information for real-time analytics.

- **Challenges:** Edge caching has a lot of advantages, but there are also a lot of difficulties. Keeping the cached data consistent with the original source data is a significant task. When the source data is regularly changed, this might be more challenging.

  Cache partitioning, or choosing how to distribute cache space among various users or categories of data, is another difficulty. The caching speed and user experience can be significantly impacted by an equitable and effective partitioning approach. Concerns about security and privacy are also quite important. To avoid unwanted access or data breaches, sensitive data kept in edge caches needs to be appropriately protected.

6.2.3   Network Optimization

Edge computing requires effective network optimization. Because edge devices are distributed geographically and require effective communication, network optimization can greatly improve system performance, lower latency, and boost user satisfaction. Edge computing network optimization encompasses a variety of strategies and factors, including load balancing, bandwidth control, and quality of service.

- **Load balancing:** A process of dividing network traffic among numerous servers or channels to make sure that no single node is overloaded is called load balancing. This maximizes throughput, minimizes delay, and makes the most use of available resources. Effective load balancing can avoid network congestion and guarantee equitable usage of network resources in the realm of edge computing. Consider an IoT device that is being served by a network of edge servers for a smart city application. A load balancer could transfer incoming requests to other less busy servers if one server becomes overloaded with traffic, maintaining dependable and effective network performance.
- **Bandwidth management:** Effective bandwidth management is essential given the rising need for high-data-rate applications like video streaming, online gaming, and real-time analytics. Service providers can improve service quality, avoid network congestion, and make the most of their network resources by managing the available bandwidth sensibly. Traffic shaping, in which the service provider regulates the rate of traffic transmitted to the network, is a typical illustration of bandwidth management in edge computing. It guarantees that key applications continue to operate without interruption during periods of high consumption, and less important applications could be given less bandwidth.
- **Quality of service:** A data flow's performance concerning priority, bandwidth, latency, and packet loss is assured by the quality of service. Network managers

can prioritize some types of traffic over others by adopting QoS methods, ensuring that crucial or latency-sensitive jobs receive the required network resources. To provide fast, real-time communication between doctors and patients even in the presence of heavy network traffic, QoS settings might, for example, prioritize video conferencing traffic in online healthcare systems.

6.2.4 Resource Allocation

In edge computing, resource allocation is a crucial aspect of optimization that involves dividing up and managing computational assets such as CPU cycles, memory, storage, and network bandwidth among various activities and services. Because resources in edge computing contexts are few and diverse, smart resource allocation may boost system performance, enhance user satisfaction, and guarantee resource utilization. In edge computing, resource allocation approaches, scheduling algorithms, and solutions for problems like resource contention and heterogeneity are important components.

- **Resource allocation approaches:** There are many different resource allocation techniques that may be used, and the choice of approach frequently relies on the particular application requirements and task characteristics. Simple techniques like "first come, first served" can be used, as well as more advanced ones that take work priorities, resource availability, user preferences, or even employ predictive analytics into consideration. An edge server may assign more processing resources to real-time danger detection jobs than to normal data storage and administration operations, similar to how this would be done in a video surveillance system because the former is given more priority.
- **Scheduling algorithms:** The timing and distribution of tasks among resources are controlled by scheduling algorithms. These algorithms take into account a number of variables, including job dependencies, resource availability, task priority, and deadlines to produce an effective timetable. Round Robin, Priority Scheduling, and Shortest Job Next are a few typical examples of scheduling algorithms. In case of a smart manufacturing setting where numerous IoT devices are gathering data and necessitating the execution of various analytics functions at the edge. An effective scheduling algorithm would make sure that these jobs are completed in a way that respects dependencies, adhere to deadlines, and makes the most use of the resources at hand.
- **Challenges:** There are several challenges with resource allocation in edge computing. Resource contention, when several activities fight for scarce resources, is a significant problem. Fairness regulations and congestion management techniques can be used to control this.

  The heterogeneity or diversity of resources is another issue. There are distinct types of edge servers, each with a varied set of resource capabilities, from highly competent micro data centers to less capable components like routers or base stations. It is difficult to create resource allocation techniques that can accommodate such variety.

6.2.5   Privacy Aware Optimization

Privacy-aware optimization has emerged as a major challenge in a constantly evolving world where edge computing devices manage significant volumes of sensitive and private data. Designing and putting into practice techniques and algorithms that optimize system performance while simultaneously preserving user privacy is known as privacy-aware optimization. It entails safeguarding sensitive data, making sure privacy rules and regulations are followed, and boosting user confidence. Privacy-preserving methods, privacy-aware algorithms, and addressing issues like the privacy-performance trade-off and regulatory compliance are important concepts of privacy-aware optimization.

- **Privacy preservation techniques:** Sensitive data may be protected in edge computing environments by using methods like anonymization, data masking, and differential privacy. An edge server, for instance, may employ anonymization techniques in a smart city application to eliminate personally identifying information from acquired data before processing it, preserving user privacy.
- **Privacy aware algorithms:** It is possible to construct algorithms and models with privacy considerations in mind. In particular, machine learning models that are sensitive to privacy might derive knowledge from data without directly viewing the original data. One such method is federated learning, in which learning occurs on the edge device itself, and only model updates are exchanged, keeping sensitive data local to the device. A personalized recommendation system for a retail business might serve as a prime example of this. Without sending private user data to a central server, edge devices may employ federated learning to build customized models.
- **Challenges:** The decision of whether to prioritize privacy or performance is one of the main challenges facing privacy-aware optimization. Performance can be negatively impacted by privacy-enhancing methods that add computational burden and delay, such as data encryption or sophisticated privacy-aware algorithms.

    Compliance with regulations is an additional challenge. It can be challenging to ensure that edge computing systems adhere to privacy laws and regulations in various locations while simultaneously optimizing performance (for example, GDPR in Europe).

    Another important factor is security. To guarantee the efficient protection of sensitive data, privacy-preserving methods and privacy-aware algorithms must be resistant to a variety of cyber threats.

6.3 Edge Optimization for Heterogeneous Devices

Edge computing environments often consist of heterogeneous devices including a vast variety of processing, storage, energy, and network access capabilities. The IoT includes a variety of devices, from robust edge servers to less capable sensors, wearables, and smart home appliances. In order to ensure the whole system operates efficiently, edge optimization must take this heterogeneity into account. Optimization primarily seeks to distribute tasks and resources across devices in the setting of heterogeneous edge computing in a way that maximizes overall system performance while lessening resource use and time to respond. The optimization may involve resource allocation, job offloading, and network optimization techniques modified to take device heterogeneity into consideration.

Let's use the example of a smart city, which often consists of a variety of gadgets, such as potent edge servers, security cameras, weather sensors, and individual mobile devices. Each gadget is different in its capabilities and qualities. A security camera may provide high-resolution video but has limited processing capacity, whereas an edge server has significant processing power but is constrained by the volume of data it can handle at once. Task offloading can be used in this situation to improve system performance. The edge servers, which have enough computational power to accomplish such jobs, may take on high-resolution video processing duties from the security cameras. Real-time video analysis is thus made possible by this offloading, which is advantageous for quick replies in traffic management or security warnings.

It's important to properly optimize resource allocation in a diverse setting. Different workloads might be assigned to various devices depending on the urgency and computing complexity of activities. Less computationally demanding jobs, like tracking ambient temperature from environmental sensors, may be done on less potent IoT gateways, leaving more resource intensive tasks to the edge servers. Another important factor is network optimization. Data transmission plans and pathways should be optimized to prevent network congestion and guarantee timely data delivery because different devices may have varying network access and bandwidth capacities. For example, the transmission of temperature data, which is less time-sensitive, should be prioritized above real-time video feeds from security cameras.

Edge optimization may make sure that each device functions within its capability boundaries while boosting the overall performance of the edge computing system by taking the heterogeneity of devices into account. Additionally, optimization can result in substantial enhancements in system responsiveness, efficiency, and energy conservation, all of which are essential for important large-scale applications like smart cities, healthcare, and industrial automation.

6.4 Edge Node Clustering

The edge node clustering technique is used in edge computing to improve resource utilization, load balancing, fault tolerance, and scalability. Edge nodes are grouped in a cluster when they have identical characteristics or are close enough to one another to function together as a single logical system. Edge node clustering has several aspects, and the complexity of its implementation depends on the particular needs of the application. Resource sharing, load balancing, fault tolerance, and node proximity are important factors that affect edge node clustering.

- **Proximity based clustering:** To bring closer data processing to users, edge nodes are frequently deployed geographically. However, more than one edge node may exist in a certain region, as may be the case in an urban setting with a high data demand. Based on their closeness, these nodes may be organized into clusters that enable more effective data sharing and coordination amongst the nodes. The user experience is improved by this method's decreased latency and faster data processing. In this case, neighboring edge nodes or data collection points in a smart city application can be clustered to process data from a specific area, enabling effective local data analysis and decision-making.
- **Load balancing:** Clustering the edge nodes can be very helpful for load balancing. Due to the fluctuating data demand, certain edge nodes may experience overloading while others go unutilized. Workloads can be divided across the nodes of a cluster to ensure a more equitably distributed utilization of resources. By avoiding misuse, load balancing not only improves system performance but also increases the lifespan of edge nodes. For instance, spectator demand for a popular event can spike on a video streaming service. All viewers will experience flawless streaming thanks to the use of edge node clustering, which distributes the burden among several nodes.
- **Fault tolerance:** System fault tolerance and reliability can both be improved by clustering. Tasks from an edge node that fails or becomes unreachable can be transferred to other nodes in the cluster to maintain service availability. For mission-critical applications where service disruption might result in large losses, this capability is very important. For example, an edge node may offer real-time navigation services in autonomous automobile systems. If this node fails, another node in the cluster will be able to take over right away, maintaining uninterrupted navigation assistance.
- **Resource sharing:** Clusters of edge nodes might also make it easier for nodes to share resources. Storage, computing power, and network bandwidth may all be pooled together and made available to the cluster as a whole. This strategy improves overall resource utilization and enables more complicated activities that a single node would be unable to complete on its own. As an example, a single-edge node may find it challenging to train a sophisticated model in a machine-learning application. However, by creating a cluster, several nodes may train the model together while using their combined resources.

6.5 Evaluation and Performance Analysis

The performance analysis and evaluation are done in order to ensure the efficacy of edge computing systems. In this process, the effectiveness, dependability, and functionality of the system are assessed using a variety of tools and approaches. The common approaches include profiling, benchmarking, tracing, containerization, and edge analytics.

6.5.1 Profiling Tools

In performance analysis and optimization, profiling tools are essential tools. They provide insights into how resources are being used, where bottlenecks exist, and how performance may be improved, which helps to understand how a system behaves. In edge computing, when resources are limited and performance standards are high, profiling is very important. The profiling tools' types, role of the tools and some associated challenges are being discussed here.

- **Types of profiling tools:** A wide variety of tools are available for profiling, from low-level ones that give thorough details on CPU usage, memory access patterns, and network activity to high-level ones that offer perceptions of the behaviour of applications and services. Low-level tools, like Linux's perf, may provide a thorough insight into CPU use and system performance. This can assist in locating CPU-intensive jobs or operations that might be optimized or delegated to other edge nodes. High-level tools, on the other hand, could concentrate on certain services or applications. In order to discover and improve slow performance or inefficient queries, a database profiling tool may offer insights into query performance.
- **Role of profiling tools:** In the optimization of edge computing, profiling tools have numerous uses. They support the optimization process by assisting in the identification of performance bottlenecks in the system, understanding resource utilisation, and validating optimization tactics. For example, a profiling tool could determine that the video encoding procedure is the key bottleneck in a scenario involving edge-based video analytics. By using this knowledge, developers might concentrate their optimization efforts on this particular area, either by implementing hardware acceleration or improving the encoding process.
- **Challenges:** There are numerous issues and challenges of profiling in edge computing. The variety of edge devices and the heterogeneity of edge workloads make it difficult to locate the right tools that can deliver accurate and meaningful profiling data. Another associated issue with profiling is how intrusive it can be. Profiling frequently entails gathering thorough data on system behaviour, which may have an impact on user experience and system performance. Therefore, it's crucial to make sure that profiling is done in a method that has the fewest possible effects on system performance.

6.5.2 Benchmarking

The practice of benchmarking revolves around the evaluation and comparison of a system's or component's performance, either against predefined metrics or in comparison to similar systems. To effectively build and implement edge computing systems, it is essential to understand the performance characteristics of various edge devices and solutions through benchmarking. Benchmarking metrics, benchmarking strategies, and benchmarking problems are important features of benchmarking in edge computing.

- **Benchmarking methods:** Different techniques may be used for edge computing benchmarking. Synthetic benchmarks involve planned workloads, such as CPU-, memory-, or network-intensive tasks, that are intended to replicate particular system behaviours. On the other hand, application benchmarks gauge performance using actual apps and workloads, such as actual video feeds and analytics jobs might be used to assess an edge computing system created for a video surveillance system.
- **Benchmarking metrics:** Performance in edge computing may be compared and measured using a variety of criteria. Latency, throughput, resource utilisation (CPU, memory, network), and power consumption are typical measures. For instance, the latency (the time it consumes to process and communicate health data) and power consumption (battery life) of an edge device created for a remote health monitoring system may be benchmarked.
- **Challenges:** Edge computing benchmarking has a number of challenges. The diversity and variability of edge devices and workloads make it challenging to create representative and equitable benchmarks. For example, a benchmark made for a strong edge server would not be appropriate for a base station or router, which are weaker edge devices. A fundamental problem in benchmarking is assuring repeatability. It might be challenging to compare benchmarking results fairly and accurately because of variations in network circumstances, user behaviour, and other variables.

6.5.3 Tracing

Tracing is a method for gathering comprehensive data on how tasks or processes are carried out in an edge computing system. It is used in the context of performance analysis and optimisation in edge computing. Tracing offers a historical description of what happened, which makes it useful for understanding the order of processes, spotting bottlenecks, resolving problems, and improving performance. The types of tracing, its role, and associated challenges are briefly discussed here.

- **Types of tracing:** There are two types of tracing in edge computing: low-level and high-level. The process of following the execution of specific instructions or system calls is referred to as low-level tracing, sometimes known as kernel tracing. This may be helpful in locating kernel- or system-level performance problems. Linux's ftrace is an example of a tool that offers low-level tracing.

High-level tracing, on the other hand, monitors the effectiveness of activities or tasks at the application level. Understanding the behaviour of particular apps or services, locating application-level bottlenecks, and improving application performance may all be facilitated by this. Zipkin and Jaeger are two tools that offer high-level tracing.

- **Role:** In performance analysis and optimization, tracing is crucial. It gives a thorough overview of how the system functions, making it easier to see how resources are allocated, where time is spent, and where possible problems might arise. For instance, tracing may show that a substantial amount of latency is being caused by the time required for data transmission between edge nodes in a scenario involving edge-based video analytics, suggesting a possible area for optimization.
- **Challenges:** The challenges of tracing in edge computing are numerous. Especially with low-level tracing, the sheer amount of data that may be produced can be overwhelming, making it challenging to analyze and glean insightful information.

Another challenge with tracing is how invasive it is. Trace data collection can have an impact on system performance and user experience, particularly in edge situations with limited resources. Hence, it is critical to achieve harmony between the necessity for thorough tracing data and any potential effects on system performance.

6.5.4 Containerization

A programme and its dependencies are packaged into a standalone, executable package called a container as part of the lightweight virtualization technique known as containerization. Containerization enhances speed, scalability, and manageability in the domain of edge computing by providing a consistent, isolated environment for executing programmes across various edge nodes. The advantages of containerization, popular tools for containerization, and challenges associated with it are discussed here briefly.

- **Benefits:** In edge computing, containerization offers a number of advantages. First off, it improves speed since containers use the same kernel as the host system and doesn't need an entire operating system, which lowers overhead. Second, it increases scalability by making it simple to start, stop, and duplicate containers across several edge nodes. Thirdly, it simplifies management since developers just need to create a single container to run an application on each edge node that supports containerization. For example, an application for processing in-store video feeds may be bundled as a container and installed on edge servers at each shop in an edge computing scenario comprising a chain of retail establishments. By processing video streams locally, this would enhance speed while also making it easier to maintain and deploy the programme across many stores.

- **Tools:** In edge computing, a variety of tools may be used for containerization. Due to its extensive feature set and broad community support, Docker is one of the most widely used technologies. Even though it isn't a solution for containerization, Kubernetes is a platform for managing and orchestrating containers, especially in large-scale and dispersed edge computing settings.

- **Challenges:** There are various challenges associated with containerization in edge computing. One of them is ensuring security since flaws in container images or the shared kernel might possibly be exploited. It might be challenging to manage and orchestrate containers across a large number of diverse edge nodes.

    The limited resource capabilities of some edge devices present another challenge. While less resource-intensive than conventional virtual machines, containers nevertheless need more resources than executing a program directly on the hardware. On less competent edge devices, this may make containerization less feasible.

6.5.5 Edge Analytics

The process of data analysis and decision-making at the edge of the network, closer to the data source, is known as edge analytics. This method significantly affects system performance optimization, reaction times, and network load in the context of edge computing. For applications that are time-sensitive or demand a high level of privacy and security, it enables real-time insights and prompt action. The edge analytics benefits, implementation methods, and challenges are briefly described here.

- **Edge analytics benefits:** The primary benefit of edge analytics is a decrease in latency. Rapid response times are achieved by doing data analysis on the edge node where it is created, eliminating the compulsion to transmit data to a central server or cloud for processing. For example, using edge analytics, an autonomous car can evaluate sensor data and make driving decisions in real time, which is essential for operating safety.

    Due to the fact that only information or insights that have been processed need to be transmitted back to the cloud or a central server, edge analytics also lessens the load on the network. In situations like industrial IoT, where copious volumes of sensor data are produced, this method is useful. Since sensitive data may be analyzed and used locally rather than being transferred and stored elsewhere, there are additional benefits, including increased privacy and security.

- **Implementation methods:** Utilizing specialized software platforms and tools made for edge computing settings, edge analytics may be deployed. For instance,

APIs are available for creating edge analytics apps on open-source platforms like Apache Edgent. Additionally, machine learning frameworks for edge devices, such as TensorFlow Lite and PyTorch Mobile, allow for the local execution of sophisticated analytical models.

- **Challenges:** Edge analytics implementation presents a number of challenges. The degree of sophistication of the analytics activities that may be completed by edge devices may be constrained by their frequent lack of processing power, storage, and battery life. Additionally, it might be challenging to manage and maintain analytics models across plenty of edge devices. To keep the models performing well and remaining accurate, they must be updated, trained, and optimized often. This is particularly difficult in expansive, scattered-edge situations.

Chapter 7
Future Work

Abstract In the "Future Work" section of this book, we traverse unexplored realms of edge computing, navigating complex challenges and emerging opportunities on the horizon. Our exploration is all-encompassing, including Cooperative Edge Computing (CEC), the optimization of wireless edge capabilities, and the fortification of edge solutions for robustness in challenging environments. Within these domains, we grapple with a multitude of complexities, encompassing security considerations, scalability impediments, performance refinement, and the integration of AI methods. This chapter also analyzes the adaptability of edge computing in Denied, Degraded, Intermittent, or Limited (D-DIL) scenarios, providing enlightenment into compatibility, prior research endeavors, and unexplored research horizons. As we project into the future, this chapter issues a call to researchers, innovators, and practitioners, urging them to initiate a quest for new knowledge, transforming the field of edge computing, and pushing the boundaries of what is doable.

Keywords Cooperative edge computing · Wireless optimization · Edge robustness · Scalability challenges · AI Integration · D-DIL environments · Performance refinement

7.1 Challenges in New Cooperative Edge Computing (CEC)

A revolutionary concept called CEC makes use of edge devices to enable collaboration and data as well as computational resource sharing for a variety of applications. Edge computing has many advantages, but there are also several significant challenges that need to be handled. A joint approach that incorporates research in computer science, networking, cybersecurity, and policy-making is necessary to address these issues in the new CEC. With the continuous evolution of this field, inventive approaches will arise, beating these constraints and facilitating the efficient and secure coordination of edge devices in extensive variety of applications. The following section examines many research problems and offers some partial solutions based on cutting-edge technologies like blockchain, DL [1], and decentralized services.

1. **Mobility challenges:** Application mobility is an important aspect to take into account. It increases application versatility while at the same time creating new problems. When users change their locations, mobile applications may encounter a fluid transition of computational resources. This ongoing process of resource movement involves the relocation of the currently active service to another device [2]. Below, we outline the primary issues associated with service migration:

   - The unpredictable nature of user mobility and request patterns makes it more difficult to implement the best service migration strategy. The dynamic influence of user equipment mobility on server performance becomes obvious when user equipment moves within a defined area and the borders of two edge servers are near together due to certain movement patterns [3].
   - When migrating services, migration duration, and cost must be carefully balanced. Ensuring there is as little latency as possible during service migration is crucial for certain latency-sensitive applications. As a result, a successful service migration strategy should aim to cut down on overall migration time. Additionally, app developers are primarily concerned with the final revenue result. The pros and cons of moving include a complex trade-off when making the decision to move. It is quite difficult to provide a migration strategy that can minimize migration costs.
   - A crucial challenge is choosing the best edge server for service migration. It's possible for the service coverage zones of many edge servers to overlap. As a result, a user must choose which server to utilize for service migration when they reach an area that is covered by many servers. Additionally, it is essential to consider the target edge server's resource availability. When moving a user's current service to the other edge server, the receiving server needs to have enough resources to efficiently handle the user's service requests.

   The MDP (Markov decision process), has already been proposed to facilitate effective decision-making in service relocation or migration. However, the comprehensive optimal solutions derived from MDP rely on simplified theoretical assumptions, that partially capture the complexities stemming from real-world conditions and a multitude of Heterogeneous attributes. Therefore, the effectiveness of these mathematical models in service migration is limited.

   Recent advancements in AI technology, particularly Deep Learning (DL), offer a promising alternative for making service migration options that consider intricate aspects such as the heterogeneity of node equipment, the dynamic nature of the network environment, and the real-time demands of user mobility. DL technology, particularly Reinforcement Learning (RL), has the ability to continuously learn from extensive historical data, adapt to ever-changing environments, and acknowledge swift changes [4]. RL can perceive its environment, take appropriate actions, and discover optimal approaches to maximize rewards in various scenarios. Within this context, three fundamental elements come into play: 'state,' which characterizes the current state of MEC servers that cover user equipment; 'action,' which represents a list of available servers for potential

VM transfers; and 'reward,' which signifies the reward a mobile user receives based on the chosen action. The overarching objective is to minimize the value of an objective function, and the Q-learning algorithm is harnessed to maximize migration rewards. Consequently, actions that result in cut communication and migration overhead for user equipment within the MEC network yield higher "rewards."

2.. **Implementation Challenges:** Deploying edge computing nodes continues to be associated with problems, such as choosing a profitable investment and operational architecture. Microservices may significantly impact how easily application packages are deployed.

- **Enterprise demands and essentials:** The most pressing issue in the deployment arena is the choice of business requirements and scenarios, which is crucial, especially in the context of 5G. Irrespective of whether we're talking about implementing edge computing for distinct clients in the enhanced mobile broadband (eMBB) context or for various business verticals like live gaming, the Internet of Vehicles, and smart manufacturing, it's mandatory to carefully evaluate the capability and viability of the particular business situation during deployment [5, 6].

- **Networking metric and financial returns:** When analyzing the stakeholders in the edge ecosystem, we can distinguish between two primary groups, infrastructure holders and software developers. Infrastructure owners, which typically include operators and leading cloud providers like Amazon and Google Cloud, make up the first category. They monitor the gathering and retention of data while also supervising the upkeep and administration of software and hardware infrastructure. This corporate model is user-centric, with members exclusively required to subscribe to the service, without the need to be familiar with the technical intricacies. The latter primarily consists of content providers and startup firms. They contribute to the rollout of edge servers, develop value-added applications, and play a vital role in enhancing and broadening innovative services. It's difficult to maintain technology that doesn't yield economic advantages. In an edge computing environment, the upkeep of hardware and software becomes especially demanding due to the dispersed nature of edge nodes. Deliberating who bears the burden of maintenance and management expenses, whether it's the cloud service provider or the content provider, is a critical factor. Additionally, achieving efficient cost reduction for users' network usage is imperative. The proximity of computing resources to the network's edge significantly enhances the user experience. However, this proximity can result in a decline in the number of access users, a drop in revenue of the edge network, and an overall increase in total costs.

- **Business framework and administration:** In edge computing, you may come across counterparts to the well-known Infrastructure as a Service (IaaS), Platform as a Service (PaaS), and Software as a Service (SaaS) from cloud computing. Operators have the capability to deliver a range of services,

including local offloading services, edge computing room rentals, and integrated IaaS functionalities tailored to the needs of diverse corporate clients. In
contrast to large enterprises, operators take a unified approach to planning and
deploying IaaS and PaaS platforms for small to medium-sized businesses, a
strategy particularly relevant when edge node resource management yields
are limited. Nevertheless, there is a possibility for extended research and
enhancement in the realm of third-party PaaS platforms and the regulation
of third-party software within the edge system [7].

- **Integrity assurance:** Preserving the safety of the physical surrounding edge
  nodes is also a challenge, primarily because there are no thorough protections
  available, including the lack of robust data backup, data recovery, and auditing
  mechanisms [8]. In contrast to the security provided by a stable cloud
  computing infrastructure, attackers could potentially alter or erase users'
  data stored on edge nodes, thereby removing crucial evidence. Therefore,
  in the development of the complete edge system, it is mandatory to utilize
  infrastructure coordination that assures physical stability and implements
  several backup measures to maintain the reliability of data.
- **Software tools:** Virtualization technologies, such as containers, are designed
  to efficiently distribute packaged applications as lightweight virtual machines
  to edge servers [9]. Even, the task of decomposing complex cloud applications
  into distributed packages and integrating them into hierarchical IoT system
  architectures, specifically to cater to application requirements like Quality of
  Service (QoS) and performance, involves substantial challenges. Resultantly,
  there arises a need to present new programming frameworks that seamlessly
  unify various facets of IoT system architecture layers to streamline distributed
  software development. Microservices emerge as a promising strategy for
  breaking down applications and services into smaller, manageable units at
  the process level. In a microservices architecture, an application is broken
  down into independent atomic services, each with minimal resource demands,
  facilitating rapid development. These service units may be operated, updated,
  and deployed autonomously, empowering development teams to achieve
  continuous delivery of functionality. Larger applications may divided into
  multiple smaller modules and deployed on edge nodes. Each module may
  effectively use computing and network resources without impacting other
  modules. The integration of distributed IoT and microservices offers a
  pathway to streamline package deployment, enhancing service delivery and
  addressing deployment hurdles tied to application packages [10].

3. **Security and confidentiality protection:** Edge computing's localized data
   processing approach minimizes the threat of user privacy during data transfer
   compared to cloud computing. Nevertheless, it comes up with unique security
   and privacy challenges stemming from the increasing presence relating to the
   heterogeneity of devices and diverse classes within edge networks.

   - **Acquire confidential information:** Edge or local computing devices are
     typically located at a short distance from the end user. Consequently, MEC

nodes situated near the user have the potential to amass sensitive data, such as the identity of users, location details, and application usage patterns [11, 12]. For example, a healthcare application is connected to an edge computing network. In this scenario, an adversary might exploit the system to monitor patterns of medical device usage, potentially compromising the privacy related to the health of patients. Additionally, the decentralized nature of MEC nodes poses a significant challenge to centralized control, which makes it exceedingly complex to manage such discrete nodes effectively.

- **Challenge in Adapting Traditional Security Measures:** In edge infrastructures, the application of conventional security and privacy protection methods, inclusively certificates, and public key infrastructure (PKI) validation, might offer certain challenges [13]. These challenges arise due to the unique nature of edge environments, which requires a reevaluation of established security practices. During the dynamic evolution of Multi-Access Edge Computing nodes, it is essential for nodes to engage in mutual verification while establishing a newly configured MEC network. Additionally, MEC nodes must also exercise vigilance in controlling service requests, ensuring that malicious or compromised nodes are prevented from network access.

- **Enhancing Device Communication Security:** In MEC networks, device communication primarily incorporates the interactions between IoT devices and MEC nodes, as well as intra-MEC node communication. Initially, terminal equipment has direct communication access to a MEC node. Nevertheless, IoT devices may remain unaware of the presence of the MEC network, rendering symmetric encryption technology unviable for securing their transmitted messages. Similarly, the constraints of asymmetric key cryptography technology may also apply. Since MEC nodes may traverse multiple paths, complete trustworthiness cannot be assured, which emphasizes the importance of incorporating end-to-end security mechanisms for communication among MEC nodes.

- **Enhancing Mobile Users' QoS while preserving Personal Information:** Service allocating within MEC is a critical research domain, with the central focus of devising an optimal method to enhance the QoS for users. Current service deployment approaches primarily rely on assessing the extent to which customers prioritize the delivery of services. However, determining customer preferences commonly requires administering sensitive personal information, including historical data and geographic locations with personalized requirements. For this reason, the implementation of a robust privacy preservation framework exhibits a considerable challenge.

The unceasingly growing variety of edge services is introducing innovative requirements for extensive privacy protection. Aside from formulating an effective privacy preservation technique, it is essential to integrate conventional privacy measures with the complications of edge data analytics in the diverse services scenario.

- **Assimilating Federated Learning (FL) with Edge Computing:** In the realm of conventional machine learning approaches, the demand to centralize training data becomes apparent, either by consolidating it onto a machine or within a cloud center. Comparatively, FL, operating as distributed DL technology, eases users to collectively train an algorithm, whereas conserving the samples of local data on respective devices [14–16]. This approach leverages a wide array of data, ranging from UE metrics such as quality of the wireless channel, battery-life, and energy consumption to insights from edge nodes comprising computing load, quality of wireless communication, and task queue. FL addresses the topmost issue of privacy leakage by evading the requirement to upload sensitive information to a centralized cloud center, instead of transmitting just the learned model weights for updates [17]. To cope with the intricacies of service deployment in MEC, FL has the ability for users to transmit their trained results without the requirement of uploading their entire dataset, including preference information, to the cloud center. This approach acts as a resilient defense by ensuring the well-protected preservation of user's private information.
- **Blockchain technology:** The decentralized ledger technology, blockchain, operates independently from a central authority and requires robust encryption techniques. It brings into being a secure, transparent, and immutable foundation for network communication, data sharing, and transactions. It also ensures the automated enforcement of some predefined rules and conditions through smart contracts while preserving data privacy and account security through asymmetric encryption algorithms [18].

  In the context of edge computing, "blockchain" signifies the capacity of network participants to record transactions within a decentralized accounting system. The principal elements of blockchain, including consensus protocols, ledger structures, and incentives, may be expanded within unified systems to cater the various tiers of edge computing environments. This integration will encompass the foundational layers of blockchain and the significant functionalities of edge computing [19]. The combination may offer enhanced security for large-scale data storage and efficient computing by eliminating the requirement of expensive encryption overheads.

7.2 Optimization in Wireless Edge Computing Environment

Wireless edge computing, commonly referred to as edge computing, utilizes the use of close proximity of computation and data processing where it originates, to lower latency and enables real-time decision-making. Optimization is an important aspect of research and development here due to the distinctive challenges and possible solutions of this closeness. This section probes into the complex area of optimization in wireless edge computing systems by exploring the main issues, recent developments, and potential future approaches.

1. **Decentralized resource allocation management:** The multi-dimensional resource-management approaches are necessary to address the dynamic behaviors of the MEC system resulting from device mobility and evolving computing application requirements. The implementation of multi-objective resource allocation approaches must be integrated with multi-dimensional schedulers. It may present a challenge due to the diverse application categories, MEC server infrastructure, varying user demands, and distinct communication channel QoS prerequisites. The wireless channel is also susceptible to congestion as the mobile device count rises, escalating competitive pressure for the limited computing resources available. While the centralized approach may settle this issue, it comes with a significant drawback of high computational complexity and substantial reporting overhead. So, the centralized process is not suitable for distributed MEC systems. Therefore, there is a need to explore dependable and distributed MEC resource allocation strategies [20].

2. **Cross platform compatibility:** In consideration of the location of users and specific technical needs, physical nodes may be strategically employed across various sites within the MEC infrastructure. A pivotal challenge is to ensure the seamless integration of MEC into the current infrastructure and interfaces while preserving the normal specifications of the core network and end devices. The capability of the MEC network to uninterruptedly share information with other elements in the 5G network is a fundamental aspect of MEC integration. It plays the primary role in workload management and the attainment of essential control information for users. There's an imperative need for application portability, which eliminates the essentiality for software developers to create diverse situations for different MEC systems [21]. In view of user locations and specific technical criteria, physical nodes may be schematically positioned within the MEC infrastructure. Thereupon a significant challenge regarding the smooth MEC integration into the existing infrastructure may arise.

3. **Reducing latency:** While EC has dominance in terms of low latency compared to cloud computing, optimization prospects are yet to be fully explored. Addressing latency challenges in edge computing is a multi-aspect task that demands a combination of network enhancements, infrastructure enhancements, and smart data handling strategies. The efforts will remain ongoing as technology transforms to fulfill the requirements of low-latency applications.

4. **Energy efficiency:** Edge computing presents an essential frontier in the exploration of efficient energy optimization, indicated by numerous unresolved dilemmas and promising research routes. The main obstacle is the limited resources of edge devices, which require creative approaches to lower energy consumption while upholding outstanding performance. Adaptive algorithms are necessary to efficiently manage resources dealing with diverse hardware and software environments at edge nodes. The real-time and dynamic workloads in edge applications add complexity, requiring energy-efficient approaches for task scheduling, resource allocation, and load balancing. Lowering the energy consumption during data transfer between edge devices and centralized data centers is a critical challenge that highlights the necessity for well-optimized

data routing and communication protocols. Energy-aware scheduling algorithms that are capable of effectively balancing task deadlines, qQoS criteria, and energy efficiency are leading current research endeavors. Predictive analytics and machine learning are prominent as robust resources for predicting energy requirements and skillfully handling energy usage in real time. Further, the persistent challenge is to manage interference and strengthen noise resilience in the edge environment. To address energy optimization thoroughly, we must consider the entire lifecycle of edge devices, such as from their inception to disposal, while following energy standards and regulations. These open challenges offer exciting research avenues for developing sustainable and energy-efficient edge computing solutions, which hold the potential to benefit both the environment and operational costs significantly. These challenges present fascinating research opportunities to design sustainable and energy-efficient solutions in edge computing, with the promise of significant advantages for both environmental conservation and cost efficiency.

5. **Cost optimization:** Within the domain of edge computing, achieving cost efficiency remains a persistent need and a continuous challenge. By expanding the usage of edge computing to respond to the demands for low-latency, high-bandwidth applications, organizations are encountering the challenge of balancing the benefits of distributed processing against the incurred cost. The costs entailed by deploying and maintaining edge infrastructure may have a sudden upswing, specifically in the geographically dispersed network of edge devices. These complexities should be acknowledged in future research in this domain to accelerate innovation. In the start, we require innovative, budget-friendly architectures and management frameworks that have the ability to effectively distribute and coordinate workloads across edge nodes, all while maintaining cost efficiency. Additionally, exploring energy-efficient edge computing solutions will play an essential role in reducing both operational costs and environmental implications. Also, a thorough investigation of the economic factors surrounding data storage and processing at the edge in contrast with centralized cloud facilities will help uncover the cost-efficiency of various deployment methods. Since edge computing continues to evolve, researchers should investigate the transformative nature of emerging technologies, like 5G, AI, and blockchain, which can be leveraged to optimize costs while ensuring robust performance and security at the edge.

6. **Quality of service:** In the promptly evolving environment of edge computing, maintaining excellent QoS is of supreme importance. To set forth the route for future research objectives, we must investigate a variety of promising directions:

   • **Dynamic QoS management:** Establish dynamic QoS implementations with the capability to immediately adapt service levels in response to real-time network conditions and the capabilities of edge devices, guaranteeing uniform performance in forever evolving edge environments.

- **Resource administration:** Explore advanced algorithms to administer the resources that are aimed at the optimal allocation of computational, storage, and edge network resources. Effectively managing resource utilization with the heterogeneous QoS needs of applications is a key challenge.
- **ML-based QoS Prediction:** Employ machine learning models and predictive analytics to predict QoS decline or resource congestion, facilitating proactive resource allocation and efficiency enhancement.
- **Multi-Tenant QoS Models:** Develop QoS frameworks and strategies to handle multi-tenancy issues in shared edge environments, assuring unbiased resource dispersion, separation, and preference for diverse tenant applications.
- **QoS integrated security measures:** Setting up security mechanisms that adopt QoS principles, seeking a balance between protecting edge services and ensuring peak performance, particularly in response to security risks.
- **QoS customization via network slicing:** Examine network slicing technologies aiming to provide custom network resources and QoS parameters for dedicated applications or service domains, facilitating in-depth QoS regulation.
- **Standardized QoS Metrics:** Create standardized benchmarks and assessment methods for evaluating QoS in edge computing environments, lessening the complexity of the comparison of performance and the evaluation of QoS optimization techniques.
- **Critical services edge QoS:** Put a strong emphasis on QoS assurance systems designed for critical edge applications, such as autonomous vehicles and healthcare devices, with the highest priority granted to reducing latency and ensuring reliability.
- **Legal and ethical implications:** Examine the legal and ethical challenges of QoS optimization in edge computing, particularly concerning data privacy, integrity, and compliance with regulatory frameworks.
- **Edge-driven Service Quality in IoT:** Progress QoS research to incorporate the specific traits of IoT devices located at the network edge, perceiving the broad scale and heterogeneity of IoT deployments.

7. **Security and privacy:** In the continuously evolving landscape of edge computing, it's undeniable that there are several exciting directions to explore as far as optimizing security and privacy:

- **Edge security paradigms:** Establish all-encompassing security methodologies customized to fit edge computing environments, conceding the individual complexities arising from diverse, distributed, and resource-constrained edge devices.
- **Zero trust models:** Probe and develop zero trust security schemes distinctively for edge computing, in order to limit the trust extent and elevate security resilience.
- **Edge-driven AI for detecting threats:** Explore the use of ML and AI on the edge for quick security threat detection and resolution, enabling proactive security measures.

- **Privacy-aware edge data analytics:** Develop models and techniques for executing analytics and data processing at the edge, with particular attention to protecting user privacy and meeting data integrity protocols.
- **Secure edge to cloud data transfer:** Analyze encryption solutions and reliable communication protocols for preserving data security during transit from edge devices to the cloud.
- **Edge computing with quantum resilience:** Anticipate the effects of quantum computing for edge security and explore quantum-resistant cryptography and security algorithms.
- **Blockchain integration:** Examine how to integrate blockchain technology into edge computing to enhance security, trust, and data integrity in distributed edge networks.
- **Secure edge organization:** Explore secure organization and orchestration of edge resources to suppress unauthorized entry and intensify resource allocation.
- **Multi-tenant security:** Cope with security issues in multi-tenant edge architectures, guaranteeing segregation and conservation of data and resources among tenants.
- **Interdomain security:** Investigate security approaches that may bridge multiple edge domains, such as edge-to-edge protection, to prohibit attacks from crossing edge boundaries.

7.3   Edge Computing in D-DIL Environments

7.3.1   D-DIL Environment Primer

A D-DIL environment presents unique challenges and prospects for edge computing. Considering such situations, the communication infrastructure and resources may be limited or uncertain, making traditional cloud computing less capable. Edge computing is distinguished in its ability to deal with these problems capably, powered by its close proximity to data sources and its ability to assess and evaluate data where it's generated.

This section is dedicated to discovering how edge computing is incorporated as a solution for D-DIL challenges. These situations, notable for constrained connectivity, unpredictable conditions, and often remote locations, have continuously been substantive limitations to stable data processing, analysis, and communication.

- **Let's examine a practical scenario**: considering the outcome of military operations, that is significantly impacted by the fast retrieval of essential information. Substantial research and development have been focused on the enhancement of critical information services, with a key concentration on domains like artificial intelligence, machine learning, edge computing services, and data analytics. However, the capability of these information services is firmly tied to the existence of resilient, efficient, and versatile communication systems that sustain consistent performance regardless of location or time. The networks operating at the tactical edge, continuously upholding highly mobile and dismounted forces, are frequently described as D-DIL [22]. This categorization arises due to issues like the challenge of staying connected during motion, lessening the need for fixed infrastructure, performing actions in urban and challenging territories, and managing electronic warfare offensives. These aspects bring about network characteristics comprising elevated and unpredictable latency, restricted bandwidth, diminished dependability, and intermittent linkages. Establishing versatile and streamlined communication systems is a critical need to enhance collaborative efforts between national military and global coalition allies. Moreover, these services are imperative for maxing out the benefits of diverse Emerging and Disruptive Technologies (EDTs), incorporating advanced analytics, big-data, autonomy, AI deployed at edge, and space technology applications.

The subsequent part is about to look into the potential of edge computing in fulfilling the requirements of D-DIL environments.

7.3.2   D-DIL and Edge Computing: Compatibility Review

In a time of ever-advancing technology, where reliability and adaptability are getting more attention, the convergence of D-DIL settings and edge computing emerges as an attractive area for research. We aim to explore how this cutting-edge approach can handle effectively the issues with inconsistent or constrained connectivity. This subsection is about to look into the potential of edge computing in fulfilling the requirements of D-DIL environments.

- **Resilient decentralization:** By deploying computational workloads nearer to the data source, edge computing lessens the requirement for centralized infrastructure in D-DIL situations. This foresighted approach affirms that essential operations carry on if connectivity faces denial or degradation.
- **Real-time data processing:** Edge computing strengthens low-latency data processing right at the source. In critical time-sensitive scenarios, D-DIL for example, prompt decision-making becomes a primary focus. Edge devices situated on-site execute data processing immediately, removing any time lags when transmitting it to remote data centers.
- **Bandwidth efficiency:** At the edge, data experiences local processing and filtering, confirming that only essential data is sent over the network. The conserving resources take precedence in D-DIL environments with unreliable or constrained network access.
- **Data security at source:** Edge computing's capability to process and store data locally acts as a protective layer against security threats during data transmission over unreliable networks. This enhances data security and privacy which is a fundamental aspect of D-DIL contexts.

- **Adaptive scalability:** The scalability of edge computing systems makes possible seamless adjustments adhering to variable resource limitations. In D-DIL situations, adaptability dominates by granting organizations the ability to customize the number of edge devices to specific requirements.
- **Energy efficiency:** The design of edge devices often prioritizes energy efficiency. This approach significantly contributes to increasing operational capabilities and resource conservation in D-DIL environments, where power is limited.
- **Redundancy and availability of data:** Since edge computing incorporates data redundancy which assures the continuous accessibility of essential data, even when a node encounters problems. In D-DIL environments, maintaining redundancy is important to ensure operational consistency.
- **Localized decision-making:** Local devices in edge computing obtain the competence to make real-time decisions, reducing dependence on centralized decision-making systems. This self-sufficiency is vital in D-DIL situations, demanding prompt actions.
- **Adaptation to resource constraints:** Edge computing solutions are developed to adapt continuously adjusting to resource changes. They have the capability to intelligently adapt their operations based on varying computing power, storage availability, and network circumstances, making them ideally suited for D-DIL environments in response to altering resource dynamics.

7.3.3   Existing Work and Research Challenges

As discussed in Sect. 7.3.2, edge computing is a potential solution to the challenges posed by the D-DIL environment. This technology leads to a revolution in the way we process data, deliver services, and assure the continuity and stability of critical operations. Within this perspective, this section investigates the persistent research initiatives and the diverse barriers that are directing the pathway of edge computing in D-DIL scenarios.

1. **Resource adaptation: Lack of tools to manage uncertainty:** Edge software systems must be capable of responding to dynamic tasks. Both the nature of edge environments and the constraints on edge device resources give rise to challenges in crafting adaptive mechanisms and successfully putting them into action during operational phases [23]. Uncertainty is ever-present, particularly when edge systems are located in D-DIL environments that pull in mission-disrupting threats. Balancing the architectural response to uncertainty relies on assessing the measure of indeterminacy in the operational environment and the availability of resources.

    Despite the profound research on adaptive and self-adaptive software systems, the effective outputs often lack the required level of maturity in industrial and defense edge applications [24]. Besides, these outcomes also lack specific identification of the sources of uncertainty, the depiction of variability ranges, and the adaptation of design methods as required. Thus, users bear the load of handling operational risks and operating with minimal confidence in the systems. Currently, developers are deficient in the necessary tools and approaches to construct systems that possess the flexibility to perceive, accommodate, and proficiently respond to alterations in the surroundings and security challenges.

2. **Data distribution and network reliability:** In the dynamic framework of edge environments, dependable network access and consistent bandwidth remain hard to attain. Particularly in the context of multi-disciplinary and collaboration-oriented edge environments, where various organizations and networks unite to expedite missions and the operational environment involves a blend of short and long-range networks, each recognized by distinct reliability features. Some network links may show reliability but limited bandwidth, whereas others may provide dependability only during predefined time slots and some might undergo intermittent connectivity issues. These complex networks connect to a diverse set of edge nodes, ranging from basic mobile devices to fairly advanced components like drones, and even portable computing environments containing server racks and network-attached storage. Capably managing this complex network of communication links and edge devices, and getting the maximum benefit from bandwidth at irregular intervals represents a considerable design challenge.

3. **Enabling AI and ML:** The power of AI and ML is gaining traction at the edge, accommodating diverse applications in missions aimed at enhancing situational awareness, strengthening intelligence analysis, and promoting the operation of autonomous systems. The impact of already mentioned challenges is increased when it comes to integrating AI and ML capabilities.

   • **Resource adaptation:** Embedding AI and ML elements within edge systems entails a significant computational overhead, often demanding specialized hardware like GPUs for optimal performance. So, balancing accuracy with resource consumption has been an ongoing challenge in the edge computing D-DIL environment. A potential approach to deal with the resource-utilization issue is to efficiently distribute the AI and ML tasks throughout edge nodes, demanding proficient distribution of computing resources.

   • **Data management and data structuring:** The continued functionality of AI and ML capabilities heavily depends on keeping the correct training data and the ability to enhance it with data collected at the edge. Considering the significant amount of edge data and the storage limitations of edge devices, the effective implementation of data collection and prioritization policies becomes critical. These policies figure out (a). the importance of data for retraining and what can be discarded, (b). the methodology for annotating collected data, and (c). the selection of summarizing and utilizing filtration methods to enhance bandwidth efficiency during data transmission. Expert-edge AI and ML systems succeed in the compatible integration of data labeling with data acquisition, which is challenging in D-DIL scenarios.

   • **Accelerated deployment:** As AI and ML approaches exhibit temporal drift and operate within ever-changing edge ecosystems, the imperative for recurrent retraining by utilizing edge-sourced data and redeployment surpasses that of traditional software components. Therefore, to maintain model relevance to the mission, essential tools, and infrastructure must be in place to facilitate model retraining, secure and efficient edge deployment, and closed-loop data collection, in D-DIL settings it's challenging.

References

1. Su, X., et al.: A comprehensive survey on community detection with deep learning. IEEE Trans. Neural Netw. Learn. Syst., early access, Mar. 9 (2022). https://doi.org/10.1109/TNNLS.2021. 3137396
2. Cao, K., Liu, Y., Meng, G., Sun, Q.: An overview on edge computing research. IEEE Access **8**, 85714–85728 (2020)
3. Siriwardhana, Y., Porambage, P., Liyanage, M., Ylianttila, M.: A survey on mobile augmented reality with 5G mobile edge computing: Architectures, applications, and technical aspects. IEEE Commun. Surveys Tutor. **23**(2), 1160–1192 (2021), 2nd Quart.
4. Kong, X., et al.: Spatial-temporal-cost combination based taxi driving fraud detection for collaborative Internet of Vehicles. IEEE Trans. Ind. Inf. **18**(5), 3426–3436 (2022)
5. Tolba, A.: Content accessibility preference approach for improving service optimality in Internet of Vehicles. Comput. Netw. **152**, 78–86 (2019)
6. Wang, W., et al.: Vehicle trajectory clustering based on dynamic representation learning of Internet of Vehicles. IEEE Trans. Intell. Transp. Syst. **22**(6), 3567–3576 (2021)
7. Song, H., Dautov, R., Ferry, N., Solberg, A., Fleurey, F.: Model-based fleet deployment of edge computing applications. In: Proc. 23rd ACM/IEEE Int. Conf. Model Driven Eng. Lang. Syst., pp. 132–142 (2020)
8. Okafor, K.: Dynamic reliability modeling of cyber-physical edge computing network. Int. J. Comput. Appl. **43**(7), 612–622 (2021)
9. Leppanen, T., et al.: Edge-based microservices architecture for Internet of Things: Mobility analysis case study. In: Proc. IEEE Global Commun. Conf. (GLOBECOM), pp. 1–7 (2019)
10. Guo, F., Tang, B., Tang, M., Zhao, H., Liang, W.: Microservice selection in edge-cloud collaborative environment: A deep reinforcement learning approach. In: Proc. 8th IEEE Int. Conf. Cyber Security Cloud Comput. (CSCloud) 7th IEEE Int. Conf. Edge Comput. Scalable Cloud (EdgeCom), pp. 24–29 (2021)
11. Husain, B.H., et al.: Survey on edge computing security. Int. J. Sci. Bus. **5**(3), 52–60 (2021)
12. Gao, L., Luan, T.H., Gu, B., Qu, Y., Xiang, Y.: Privacy-Preserving in Edge Computing. Springer, Singapore (2021)
13. Ranaweera, P., Jurcut, A.D., Liyanage, M.: Survey on multi-access edge computing security and privacy. IEEE Commun. Surveys Tutor. **23**(2), 1078–1124 (2021)
14. Ye, D., Yu, R., Pan, M., Han, Z.: Federated learning in vehicular edge computing: A selective model aggregation approach. IEEE Access **8**, 23920–23935 (2020)
15. Khan, L.U., Saad, W., Han, Z., Hossain, E., Hong, C.S.: Federated learning for Internet of Things: Recent advances, taxonomy, and open challenges. IEEE Commun. Surveys Tutor. **23**(3), 1759–1799 (2021), 3rd Quart.
16. Li, A., et al.: LotteryFL: Empower edge intelligence with personalized and communication-efficient federated learning. In: Proc. IEEE/ACM Symp. Edge Comput. (SEC), pp. 68–79 (2021)
17. Lim, W.Y.B., et al.: Federated learning in mobile edge networks: A comprehensive survey. IEEE Commun. Surveys Tutor. **22**(3), 2031–2063 (2020), 3rd Quart.
18. Firdaus, M., Rhee, K.-H.: On blockchain-enhanced secure data storage and sharing in vehicular edge computing networks. Appl. Sci. **11**(1), 414 (2021)

19. Yang, R., Yu, F.R., Si, P., Yang, Z., Zhang, Y.: Integrated blockchain and edge computing systems: A survey, some research issues and challenges. IEEE Commun. Surveys Tutor. **21**(2), 1508–1532 (2019), 2nd Quart.
20. Pham, Q.V., Anh, T.L., Tran, N.H., Park, B.J., Hong, C.S.: Decentralized computation offloading and resource allocation for mobileedge computing: A matching game approach. IEEE Access **6**, 75868–75885 (2018)
21. Pham, Q.-V., Fang, F., Ha, V.N., Piran, M.J., Le, M., Le, L.B., Hwang, W.-J., Ding, Z.: A survey of multi-access edge computing in 5G and beyond: Fundamentals, technology integration, and state-of-the-art. IEEE Access **8**, 116974–117017 (2020)
22. Barz, C., et al.: Enabling adaptive communications at the tactical edge. In: MILCOM 2022 - 2022 IEEE Military Communications Conference (MILCOM) (2022)
23. Ozkaya, I., Pitstick, K.: Engineering of edge software systems. Dtic.mil (2022) [Online]. Available: https://apps.dtic.mil/sti/trecms/pdf/AD1207039.pdf. Accessed 10 Sept 2023
24. Szabo, C., Sims, B., Mcatee, T., Lodge, R., Hunjet, R.: Self-adaptive software systems in contested and resource-constrained environments: Overview and challenges. IEEE Access **9**, 10711–10728 (2021)

Glossary

Active Backup A strategy to secure fault tolerance in which a redundant backup system consistently maintains and secures the latest copy of the primary system's data and functionalities, set to acquire control if the primary system runs into a malfunction.

Applications of Edge Computing Various real-life scenarios and practical instances where edge computing technologies are employed for data processing and application execution near the information source, Minimizing response time and uplifting real-time data processing.

Benchmarking The practice of assessing the performance of s system or component against the established standards or reference points.

Bottleneck Detection Uncovering factors within a system that repress its efficiency.

Causes and Types of Faults Detailed investigation of the contributing factors to potential failures or faults in edge computing systems, including hardware and software-related challenges, together with an investigation of diverse fault classifications.

Check Pointing and Scale Out A practice in fault-tolerant systems where checkpoints are carefully arranged throughout a program's execution, enabling it to come back from failures and proceed from an acknowledged state.

Containerization Packaging applications along with their dependencies into isolated, lightweight containers is an approach that significantly enhances portability and scalability in deployment.

Cooperative Edge Computing Edge devices and nodes collaborate to jointly execute computations and combine resources to attain mutual goals.

D-DIL (Denied, Disconnected, Intermittent, or Limited) Environments Scenarios where network conditions pose challenges, featuring occasional network disruptions or bandwidth limitations.

Edge Analytics Local data analysis at the network's edge, supporting real-time insights and quick decision-making.

© The Author(s), under exclusive license to Springer Nature Singapore Pte Ltd. 2024
Y. Zhai et al., *Edge Computing Resilience*, SpringerBriefs in Computer Science,
https://doi.org/10.1007/978-981-97-6998-8

Edge Computing A distributed computing approach that gives precedence to data processing and analysis at the edge of the network, lowering latency and optimizing real-time data management.

Edge Computing Resilience The capacity of edge computing systems to withstand and recover from faults, failures, or disruptions, all while upholding uninterruptible service delivery.

Edge Node A device placed at the network's edge, with the fundamental purpose of processing and managing edge computing operations.

Edge Node Clustering The method of creating clusters in edge computing, aimed at better resource management, load distribution, and strengthened fault robustness through the grouping of edge nodes or devices.

Energy Aware Offloading An approach for task offloading that includes energy-aware considerations, focusing on energy efficiency enhancement.

Edge Caching Caching frequently accessed data or content near the network's edge, designed to optimize content delivery and minimize latency.

Fault Tolerance in Edge Computing Edge computing systems' capacity to tolerate and move on functioning in the presence of hardware or software failures, maintaining a high level of continuity and trustworthiness.

Network Optimization Approaches and practices adopted to boost network performance, concentrating on latency reduction and the efficient data transit in the context of edge computing environments.

Optimization in Edge Computing Environment Elevating the performance, efficiency, and resource utilization in edge computing systems by implementing different techniques and plans.

Privacy-Preserving Offloading Scheme The process of securely distributing computational tasks to edge computing resources from local devices, and simultaneously upholding the privacy of sensitive data, implementing a fusion of approaches and procedures.

Profiling Tools The use of software tools to amass performance metrics, including CPU and memory utilization, with the aim of analyzing and fine-tuning application performance in edge computing settings.

Resource Allocation Efficient resource provisioning in an edge computing environment, incorporating the allocation of CPU, memory, and network bandwidth to different tasks or applications with the intention of optimizing efficiency.

Scalability for Distributed Edge Systems The capacity of an edge computing system to scale and adapt to rising workloads and demands through the versatile integration or reallocation of resources.

SDN (Software-Defined Networking) A network structure that encourages the centralized management of network resources, fostered by software applications with programmable capabilities.

State Backup and Scale Methods for retaining and securing the state or data in an edge computing application, with the objective of preserving data integrity and ensuring availability when failures occur.

Task Offloading The act of shifting computational tasks from local devices to high-powered edge computing resources with the purpose of enhancing both performance and efficiency.

Tracing The practice of observing and logging the execution sequence of applications operating in an edge computing environment, considering the motives of debugging, performance analysis, and optimization."